Nonlinear Control and Filtering for Stochastic Networked Systems

Nonlinear Control and Filtering for Stochastic Networked Systems

Lifeng Ma
Zidong Wang
Yuming Bo

CRC Press
Taylor & Francis Group
Boca Raton London New York

CRC Press is an imprint of the
Taylor & Francis Group, an **informa** business

Published in 2019 by CRC Press
Taylor & Francis Group
6000 Broken Sound Parkway NW, Suite 300
Boca Raton, FL 33487-2742

First issued in paperback 2020

ISBN 13: 978-0-367-65686-7 (pbk)
ISBN 13: 978-1-138-38657-0 (hbk)

**Visit the Taylor & Francis Web site at
http://www.taylorandfrancis.com**

**and the CRC Press Web site at
http://www.crcpress.com**

This book is dedicated to the Dream Dynasty, consisting of a group of bright people who have been to Brunel University London to enjoy happy research by handling both stochasticity and nonlinearity through filtering noises as well as controlling complexities in a networked way ......

Contents

Preface

The rapid development in network technologies in the past few years has led to a revolution in engineering practices. Using a network as the basement of a feedback control loop is now widely applied in various types of engineering programs, and therefore, recently, control and filtering of networked systems have been an active branch within the general research area of systems science and signal processing. The usage of the network in control systems has many advantages such as low cost, reduced weight and simple installation, as well as certain limitations (e.g., network-induced time-delays, data package loss, etc.) due mainly to complex working conditions and limited bandwidth. As is well known, in the past few decades, there have been extensive study and application of stochastic systems because the stochastic phenomenon is inevitable and cannot be avoided in the real-world systems. When modeling such kinds of systems, neglecting the stochastic disturbances, which is a conventional technique in traditional control theory for deterministic systems, is not applicable anymore. Having realized the necessity of introducing more realistic models, nowadays, a great number of real-world systems such as physical systems, financial systems, ecological systems as well as social systems are more suitable to be modeled by stochastic systems, and therefore the stochastic control/filtering problems which deal with dynamical systems, described by difference or differential equations, and subject to disturbances characterized as stochastic processes have drawn much research attention.

In this book, we discuss the recent advances in control and filtering problems for stochastic systems whose communications are based on various types of networks. The stochastic dynamics taken into consideration are quite general and could cover a wide range of well-studied nonlinear/linear and time-invariant/time-varying stochastic systems. Some sufficient conditions are derived for the existence of the desired controllers and filters by means of certain convex optimization algorithms. The compendious frame and description of the book are given as follows. Chapter 1 introduces the recent progress in control and filtering problems for stochastic systems and the outline of the book. Chapter 2 is concerned with the $\mathcal{H}_\infty$ sliding mode control problem for a class of nonlinear stochastic systems with data packets losses. In Chapter 3, the discrete sliding mode concept is adopted to investigate the control issue for a class of uncertain nonlinear networked systems with multiple communication delays. Chapter 4 discusses the robust sliding mode controller design problem for nonlinear stochastic systems subject to randomly occurring communication delays. Chapter 5 deals with the reliable $\mathcal{H}_\infty$ control problem for a class of nonlinear stochastic systems subject to possible randomly occurring sensor failures. The consensus control problem is investigated in Chapter 6 for a class of discrete time-varying stochastic multi-agent systems subject to sensor saturations based on an event-triggering mechanism. In Chapter 7, the mean-square consensus performance is taken into consideration for a type of stochastic multi-agent systems whose individual agents are forced to reach consensus over a pre-specified time interval provided the designed control protocol. Chapter 8 deals with the consensus control problem for a class of discrete time-varying stochastic multi-agent systems with stochastic nonlinearities and deception attacks. For a class of time-varying stochastic systems subject to sector nonlinearities, unknown but bounded disturbances and sensor saturations, Chapter 9 investigates the set-membership filtering problem over a finite horizon by means of an event-triggering mechanism. Chapter 10 examines the variance-constrained distributed filtering problem for a class of time-varying systems subject to multiplicative noises, unknown but bounded disturbances and deception attacks over sensor networks. Chapter 11 gives the conclusion and some possible future research topics. This book is a research monograph whose intended audience is graduate and postgraduate students as well as researchers.

Acknowledgements

We would like to acknowledge the help of many people who have been directly involved in various aspects of the research leading to this book. Special thanks go to Professor Qing-Long Han from Swinburne University of Technology, Melbourne, Australia, Dr. Hak-Keung Lam from King's College London, London, the United Kingdom, Dr. Nikos Kyriakoulis from Brunel University London, London, the United Kingdom, and Dr. Derui Ding from University of Shanghai for Science and Technology, Shanghai, China. Special thanks are given to Mr. Lei Liu, Hao Sun, Yuqi Ma and Yiwen Wang for their tremendous help in the editorial and proofreading work.

Symbols

$\mathbb{R}^n$	The n-dimensional Euclidean space.
$\mathbb{I}^+$	The set of nonnegative integers.
$\mathbb{R}^{n \times m}$	The set of all $n \times m$ real matrices.
$\| \cdot \|$	The Euclidean norm in $\mathbb{R}^n$.
$l_2[0, \infty)$	The space of square-integrable vector functions over $[0, \infty)$.
$\rho(A)$	The spectral radius of matrix A.
$\mathrm{tr}(A)$	The trace of matrix A.
$\otimes$	The Kronecker product of matrices.
$\mathrm{st}(A)$	The stack that forms a vector out of the columns of matrix A.
$\|a\|_A^2$	Equals to $a^{\mathrm{T}} A a$.
$\mathcal{P}\{\cdot\}$	The occurrence probability of the event "$\cdot$".
$\mathbb{E}\{x\}$	The expectation of stochastic variable x.
$\mathbb{E}\{x|y\}$	The expectation of x conditional on y, x and y are all stochastic variables.
I	The identity matrix of compatible dimension.
$X > Y$	The $X - Y$ is positive definite, where X and Y are symmetric matrices.
$X \geq Y$	The $X - Y$ is positive semi-definite, where X and Y are symmetric matrices.
M^{T}	The transpose matrix of M.
$\mathrm{diag}\{M_1, ..., M_n\}$	The block diagonal matrix with diagonal blocks being the matrices $M_1, ..., M_n$.
$*$	The ellipsis for terms induced by symmetry, in symmetric block matrices.

1

Introduction

1.1 Nonlinear Stochastic Networked Systems

The rapid development in network technologies in the past few years has led to a revolution in engineering practices [36]. Using a network as the basement of a feedback control loop is now widely applied in various types of engineering programs; this has gained a great deal of research interest [36, 165, 213, 255]. The usage of networks in control systems has many advantages such as low cost, reduced weight and simple installation, as well as some limitations including the network-induced time-delays (also called communication delays) and data package loss caused mainly by the complex working conditions and limited bandwidth. As a result, most of the literature on network systems has focused on how to eliminate or compensate for the effect caused by the communication delays [165, 196] and data package loss [116, 187]. Many researchers have studied the stability and controller design problems for networked systems in the presence of deterministic communication delays. Recently, due to the fact that such kinds of time-delays usually appear in a random and time-varying fashion, the communication delays have been modeled in various probabilistic ways, see [165, 196, 219, 236, 250], among which the binary random delay has gained particular research interest because of its simplicity and practicality in describing network-induced delays [219, 236].

As is well known, in the past few decades, there have been extensive study and application of stochastic systems because the stochastic phenomenon is inevitable and cannot be avoided in real-world systems. When modeling such kinds of systems, way neglecting the stochastic disturbances, which is a conventional technique in traditional control theory for deterministic systems, is not suitable anymore. Having realized the necessity of introducing more realistic models, today, a great number of real-world systems such as physical systems, financial systems, ecological systems as well as social systems are more suitable to be modeled by stochastic systems, and therefore the stochastic control problem, which deals with dynamical systems, described by difference or differential equations, and subject to disturbances characterized as stochastic processes, has drawn much research attention; see [7] and the references therein. It is worth mentioning that a kind of stochastic system represented as a deterministic system adding a stochastic disturbance characterized as white noise has gained special research interest and found extensive applications in engineering based on the fact that it is possible to generate stochastic processes with covariance functions belonging to a large class simply by sending white noise through a linear system. Hence a large class of problems can be reduced to the analysis of linear systems with white noise inputs; see [27, 33, 37, 87, 185] for examples.

Parallel to the control problems, the filtering and prediction theory for stochastic systems which aims to extract a signal from observations of signal and disturbances has been well studied and found widely applied in many engineering fields. It also plays a very important role in the solution of the stochastic optimal control problem. Research on the filtering problem was originated in [223], where the well-known Wiener-Kolmogorov filter has been proposed. However, the Wiener-Kolmogorov filtering theory has not been widely applied mainly because it requires the solution of an integral equation (the Wiener-Hopf equation) which is not easy to solve either analytically or numerically. In [94, 95], Kalman and Bucy gave a significant contribution to the filtering problem, by using the celebrated Kalman-Bucy filter which could solve the filtering problem recursively. The Kalman-Bucy filter (also known as the $\mathcal{H}_2$ filter) has been extensively adopted and widely used in many branches of stochastic control theory, since the fast development of digital computers recently; see [12, 74, 90, 130] and the references therein.

In real-world engineering, it is well acknowledged that almost all practical systems are time-varying. For such time-varying systems, a filter that could provide better transient performance than those traditional methods developed to achieve specified steady-state performance is more effective and applicable. Therefore, the filtering problems for time-varying systems have stirred considerable research interest in the past few years. For example, the difference Riccati equation method has been proposed in [237] to solve the robust Kalman filtering problem for uncertain time-varying systems. Recently, the recursive linear matrix inequality (RLMI) method has become another effective approach to deal with the filtering and control problems for time-varying systems. Originally proposed in [65], the RLMI method has so far been widely recognized and extensively utilized in both theoretical research and engineering applications associated with time-varying systems, see e. g. [40, 193]. However, up to now, the distributed filtering problem has not been adequately investigated yet for systems subject to time-varying parameters, especially for the case where the event triggering mechanism and sensor saturation are also involved.

A quintessential example that should be cited is that, up to now, most multi-agent systems (MASs) discussed in the literature have been assumed to be *time-invariant* and *deterministic*. This assumption is, however, very restrictive as almost all real-world engineering systems have certain parameters/structures which are indeed *time-varying* [11]. For such time-varying systems, a finite-horizon controller is usually desirable as it could provide better transient performance for the controlled system especially when the noise inputs are non-stationary; see [50, 65] for some recent results. However, when it comes to the consensus of multi-agent systems, the corresponding results have been scattered due mainly to the difficulty in quantifying the consensus over a finite horizon. It is notable that the consensus problem for MASs with time-varying parameters has received some initial research attention (see e.g. [92, 115, 257]). Nevertheless, the research on time-varying multi-agent systems is far from adequate and there are still many open challenging problems remaining for further investigation. On the other hand, sensor saturation is a frequently encountered phenomenon resulting from physical limitations of system components as well as the difficulties in ensuring high fidelity and timely arrival of the control and sensing signals through a possibly unreliable network of limited bandwidth. In other words, the sensor outputs are often saturated because the physical entities or processes cannot transmit energy and power with unbounded magnitude or rate. As such, it makes practical sense to take sensor saturation into account when dealing with the output-feedback control problems for time-varying MASs, which remains as an ongoing research issue.

1.2   Sliding Mode Control

Due to the advantage of strong robustness against model uncertainties, parameter variations and external disturbances, the sliding mode control (SMC) (also called variable structure control) problem of continuous-time systems has been extensively studied and widely applied in various fields; see, for example, [25, 79, 166, 167, 208, 218]. In recent years, the SMC problem for discrete-time systems has begun to receive increasing research attention simply because most control strategies are implemented in discrete time nowadays [2, 16, 24, 32, 34, 52, 69, 89, 91, 109, 212, 230, 233, 266]. In [2, 16], the integral type SMC schemes were proposed, respectively, for sample-data systems and a class of nonlinear discrete time systems. In [24, 32], adaptive laws were applied to the sliding mode control problems for discrete time systems with stochastic or deterministic disturbances. In [34], a simple methodology for designing sliding mode controllers was proposed for a class of linear multi-input discrete-time systems with matching perturbations. In [91], a discrete variable structure control method with a finite-time step to reach the switching surface was constructed by using the dead-beat control technique. In [52, 109], the discrete time SMC problems were solved via output feedback in the case when the system states are not available. It is worth mentioning that, in [212], a reaching law approach was introduced which can be conveniently used to develop the robust control law, and has therefore attracted quick attention in recent literature, see e.g. [230, 233, 266]. By employing such a reaching law and the proposed technique, in [230], the sliding mode control problem was tackled for discrete-time systems with input delays; in [266], a class of nonlinear systems was first modeled as a T-S fuzzy model and then stabilized by an SMC controller; and in [233], the SMC problem was studied for a class of uncertain discrete-time systems with time-varying but deterministic state delays.

In a networked environment, it seems sensible to apply sliding mode control strategies by taking the network-induced phenomena into account, in order to make full use of the robustness of SMC against model uncertainties, parameter variations and external disturbances. Moreover, the model under consideration should be more comprehensive to reflect the realities such as the discrete-time nature, the state-dependent stochastic disturbances (also called multiplicative noises or Itô-type noises) and the multiple randomly occurring communication delays. The possible combination of SMC and NCS would clearly have both theoretical significance and application potential. Unfortunately, to the best of the authors' knowledge, the *discrete-time* sliding mode control problem for networked systems has not been fully addressed, not to mention the consideration of stochastic delays and stochastic disturbances. This is mainly due to the mathematical challenges in analyzing the reaching condition for the addressed discrete-time stochastic systems.

1.3 Distributed Filtering

Sensor Networks

The rapid development of microelectronic technologies over the past few decades has boosted the utilization of networks which consist of a large number of devices implemented distributively for sensing, communication as well as actuating [194]. A quintessential example is the sensor network which has found wide applications ranging from various industrial branches to critical infrastructures such as military facilities and power grids; see [19,61,68,145,154,248] and the references therein. The practical significance of sensor networks has recently led to considerable research interest on the distributed estimation or filtering problems whose aim is to extract the true signals based on the information measurements collected/transmitted via sensor networks. Compared with traditional filtering algorithms in a single sensor system [106,146,197,254], the key feature of the distributed filtering over sensor networks is that each sensor estimates the system state based not only on its own measurement but also on the neighboring sensors' measurements according to the topology [206]. So far, much effort has been made to investigate the sensor network related problems and several effective strategies have been developed; see [40] for a survey. In particular, the distributed state estimation or filtering problems over sensor networks have posed many emerging challenges, which have attracted ever-increasing research attention within the signal processing and control community.

Despite the merits, it is well known that the embedded micro-processors are typically of limited capacity within a sensor network due primarily to the physical and communication constraints. Consequently, some new phenomena (e.g. signal quantization, sensor saturation and actuator failures) have inevitably emerged that deserve particular attention in system design. These phenomena are customarily referred to as the incomplete information that has attracted much research interest in developing filtering schemes [11,40,44,47,81,85,97,117,125,132,193,234,235,245,253,272]. However, when it comes to event-based distributed filtering problems with incomplete information, the corresponding results have been very few owing mainly to the lack of appropriate techniques for coping with 1) the complicated node coupling according to the topological information and 2) the demanding triggering mechanism accounting for the limited capability. As such, another motivation for our current investigation is to examine the impact of the incomplete information on the performance of the event-based distributed filtering over the sensor network with a given topology.

Distributed Filter Design Methodologies

So far, considerable effort has been devoted to the investigation of distributed filtering problems and many strategies have been developed based on the Kalman filtering theory or the $\mathcal{H}_\infty$ filtering theory; see [40] for a survey. Some representative works can be summarized as follows. By means of the Kalman filtering algorithm, a distributed filter has been designed in [172], guaranteeing that the nodes of a sensor network can track the average of all the sensors' measurements. The results have been extended in [171] to the case with complex communication modes and data-packet dropouts. For time-varying systems, the distributed finite-horizon Kalman filtering problem has been solved in [28] in the context of wireless sensor networks with limited bandwidth and energy. Apart from the Kalman filtering algorithm, the $\mathcal{H}_\infty$ filtering theory is another available technique which has been frequently utilized in filtering problems with regard to sensor networks [50,144,206]. For instance, [206] has presented a robust consensus filtering approach which ensures a suboptimal performance characterized, in an $\mathcal{H}_\infty$ sense, by the disagreement of estimates from all the sensing nodes. A similar problem has been solved in [206] for the case of switching topology. In [50], a generic framework has been established for designing the distributed filter to reach an $\mathcal{H}_\infty$-type consensus performance despite the quantization error and data-packet dropout.

As is well known, the Kalman filtering technique requires an assumption of Gaussian distributions for process and measurement noises, while the $\mathcal{H}_\infty$ theory can be utilized in the occasion when the disturbances are assumed to have bounded energy. However, in many real-world engineering practices, due to a variety of reasons (e.g., man-made electromagnetic interference), it is much more appropriate to model the disturbances/noises as signals that are unknown but bounded in certain sets rather than Gaussian noises or energy-bounded disturbances [60,107,144]. Obviously, in such a case, the afore-mentioned conventional techniques based on Kalman filtering or $\mathcal{H}_\infty$ filtering frameworks are no longer effective. Consequently, the filtering problems for systems subject to the so-called unknown but bounded (UBB) noises have excited tremendous fascination in researchers as well as engineers within the signal processing community. So far, quite a few methodologies have been exploited; see e.g. [66,151]. Nevertheless, in the general context of sensor networks, little progress has been made on the corresponding filtering problems owing probably to the difficulty in quantifying the filtering performance with respect to the unknown but bounded noises as well as the complexity which stems from the coupling between communication topology and the system dynamics.

On another research frontier, the set-membership filtering problem originated in [224] aims to use measurements to calculate recursively a bounding ellipsoid to the set of possible states; see [13,183] and the references therein. Recently, there has been renewed interest in the set-membership filtering problems for various systems by developing computationally efficient algorithms. For instance, in [66], the convex optimization method has been utilized to handle set-membership filtering with guaranteed robustness against the system parameter uncertainties. In [72], the set-membership filtering issue has been discussed in the frequency domain and an adaptive algorithm has been developed with applications in the frequency-domain equalization problem. It is worth mentioning that the set-membership filtering problem has been addressed in [235] for a stochastic system in the presence of sensor saturations, where a recursive scheme has been provided for constructing an ellipsoidal state estimation set of all states consistent with the measured output and the given noise. Unfortunately, for large-scale distributed systems such as sensor networks, the set-membership filtering has not received adequate research attention, and this motivates us to investigate the set-membership filtering problem for nonlinear systems under the distributed information processing mechanism.

1.4   Cyber Attacks

Along with the pervasive utilization of open yet unprotected communication networks, sensor networks are vulnerable to cyber threat [67,82]. As a result, the security of the network, which is of utmost importance in networked-related systems, has provoked an increasing research interest and a multitude of results have been reported in the literature; see [200] and the references therein. In general, there are mainly two types of cyber attacks which can affect the system's behavior directly or through feedback, namely, denial-of-Service (DoS) attacks [129] and deception attacks [56]. The DoS attack deteriorates system performance by preventing the information from reaching the destination, while the deception attack aims at manipulating the system toward the adversaries' desired behaviors by tampering with the control actions or system measurements. To tackle filtering/control problems for the systems under cyber attacks, several approaches have been developed, including linear programming [200], the linear matrix inequality (LMI) method [173], the game theory approach [4, 267], to name but a few key ones. However, when it comes to the distributed filtering issues over sensor networks, the corresponding results have been scattered, although some interesting initial results have appeared; see e.g. [211, 249]. To the best of the authors' knowledge, up to now, the research on distributed filtering is far from adequate when the communication networks are affected by attacks. The difficulty probably lies in the lack of appropriate attack models which, on one hand, could comprehensively reflect the engineering practice, and on the other, can be handled systematically within the existing frameworks. There are still many open yet challenging problems deserving further investigation.

On the other hand, during the process of information transmission between the agents via the communication networks, the data without security protection can be easily exploited by attackers. The network-based attacks give rise to certain challenges for those engineering applications such as consensus control for MASs since they could result in unsatisfactory consensus performance or might even cause agents to collide if not handled appropriately. Therefore, much effort has been devoted to the investigation recently of the security of networks and many results have been reported; see [5, 56, 129, 173] and the references therein. As aforementioned, in the general context of multi-agent systems, there are also mainly two types of attacks. In DoS attacks, the adversary prevents the agents from receiving information sent from their neighbors, while in deception attacks, the adversary replaces the true data exchanged between the agents by false signals [23]. So far, a slice of preliminary results concerning security problems of multi-agent systems has appeared; see [10, 14, 59, 188, 189] and the references therein. However, when it comes to the consensus control problem for MASs, especially on the occasion when the systems under investigation are time-varying, the relevant results have been significantly scattered. This is mainly because it is always arduous to quantify the consensus behavior in a time-varying manner, not to mention in the case where cyber attacks are involved.

1.5   Network-Induced Randomly Occurring Phenomena

It is widely recognized that in almost all engineering applications, nonlinearities are inevitable and could not be eliminated thoroughly. Hence, the nonlinear systems have gained more and more research attention, and lots of results have been published; see [26, 63, 104, 124, 193] for some latest results. On the other hand, due to the wide appearance of stochastic phenomena in almost every aspect of our daily life, stochastic systems which have found successful applications in many branches of science and engineering practice have stirred quite a lot of research interest during the past few decades. Therefore, the control as well as filtering problems for stochastic nonlinear systems have been studied extensively so as to meet ever-increasing demand toward systems with both stochasticity and nonlinearities; see e.g. [70, 84, 101, 135, 193, 199, 222, 240] and the references therein.

On the other hand, in real-world engineering practice, it is well known that the sensors have always been confronted with different kinds of failures due to a variety of reasons, including the erosion caused by severe circumstance, abrupt changes of working conditions, intense external disturbance and internal component constraints and aging [210, 235]. As requirements increase for reliability of engineering systems, the fault-tolerant control problem, also known as the reliable control problem which aims to stabilize the systems accurately and precisely in spite of the possible failures, has therefore attracted considerable attention in the past decades [177, 186]. In [238], an adaptive reliable $\mathcal{H}_\infty$ filter was designed with the assumption that the sensor might fail within a known bound. Such a sensor failure model is quite general, and hence has been applied in many papers to deal with reliable control as well as filtering problems; see [128, 139] for examples. It is worth pointing out that, in many real-world engineering practices, due to random abrupt changes of the environmental circumstances or working conditions, most of the time the sensors might encounter failures in a probabilistic way, that is, *randomly occurring sensor failures*. Different from the traditional deterministic sensor failure model, such randomly occurring sensor failure could be either existent or non-existent at a specific time and the existence is governed by certain Bernoulli distributed white sequences with known probabilities. However, in spite of its clear physical insight and importance in engineering application, the control problem for nonlinear stochastic systems under the circumstance of randomly occurring sensor failures has not yet been studied sufficiently.

On another research front, in most practical systems nowadays such as a target tracking system, there may be certain observations that consist of noise only when the target is absent due to its high maneuverability. In other words, the measurements are not consecutive but usually subject to partial or complete information missing. Such a phenomenon is referred to as measurement missing, information dropout or data packet losses, which occurs frequently for a variety of reasons such as sensor temporal failure, network congestion, accidental loss of some collected data or network-induced delay, and might lead to system performance degradation and sometimes even instability. Therefore, in the past few years, a great deal of research effort has been made to solve control and filtering problems in the presence of data packet losses. Such a data packet loss phenomenon is usually characterized in a probabilistic way. As a result, in [116, 183, 187, 220], the packet losses phenomenon has been modeled by different kinds of stochastic variables, among which the binary random variable sequence taking on values of 0 and 1, also known as the Bernoulli distributed model, has been widely applied because of its simplicity [220].

1.6 Nonlinear Stochastic Multi-Agent Systems

The past decade has seen a surge of research interest in multi-agent systems due primarily to their extensive applications in a variety of areas ranging from the chemistry manufacturing industry, geological exploration, building automation, to military and aerospace industries [9,51,62,73,110,133,141,142, 152,158]. MASs are comprised of a large number of interaction/cooperation units called agents which directly interact with their neighbors according to a given topology. The MASs often display rich yet complex behavior even when all the agents have tractable models and interact with their neighbors in a simple and predictable fashion [6]. Because of its clear physical and engineering insights, the consensus problem of MASs has been garnering considerable research attention and many results have been reported in the literature, such as those on cooperative control of unmanned air vehicles (UAVs) or unmanned underwater vehicles (UUVs) [178], formation control for multi-robot systems [57], collective behaviors of flocks or swarms [170], and distributed sensor networks [41,49,191,206].

One of the major lines of research on multi-agent systems is the so-called consensus control, whose objective is to design a consensus algorithm (or protocol) using local neighboring information such that the states of a team of agents reach some common features. Here, the common features are dependent on the states of all agents and examples of the features include positions, phases, velocities, attitudes, and so on. The past several years have witnessed a considerable surge of interest regarding MASs with close ties to consensus problems. These relevant research fields include but are not limited to 1) consensus which means a collective of agents (sub-systems) reach a common value that is dependent on the states of all agents [57]; 2) collective behavior of flocks and swarms, such as fish, birds and bees where a large number of agents are interacting with neighbors with a predetermined common group goal [204]; 3) sensor fusion in which a distributed filtering algorithm is adopted that allows the nodes of a sensor network to track the average of n sensor measurements using an average consensus based distributed filter called a consensus filter [172]; 4) random networks whose topology varies over time in a manner that the existence of an information channel among individual agents is random and obeys certain probabilistic distribution [76]; 5) synchronization of coupled oscillators which is recognized as a popular phenomenon in physics, engineering, biology and other relevant areas, where a large quantity of coupled agents oscillate at a common pulsation [93]; 6) algebraic connectivity of complex networks which is expressed by the second smallest eigenvalue of the Laplacian matrix of the network, characterizing the speed of convergence of the consensus algorithm [169]; 7) asynchronous distributed algorithms by which each individual agent calculates and updates at its own pace rather than the conventional mechanism where all the agents' decisions have to be synchronized to a common clock [55]; 8) formation control whose purpose is to drive a collection of mobile autonomous vehicles to accomplish the predetermined tasks while moving together according to certain required formation [205]; and 9) a cooperative control problem which deals with the cooperation among a group of interacting agents to perform a shared task by exchanging the individual agent's information to corresponding neighbors via communication networks [57].

It is worth mentioning that the fundamental problem of the above problems is the consensus problem. Generally speaking, the consensus problem aims to design a control scheme/protocol which characterizes the way in which the individual agent interacts with neighbors, thereby driving all the states of a network of agents to reach a certain common value of interest. The following gives an explicit definition on the consensus of a multi-agent system.

Definition 1.1 *Consider a multi-agent system with N agents. Denote by x_i $(i = 1, 2, 3, \ldots, N)$ the state of the i-th agent. The multi-agent system is said to reach consensus if $\forall i \neq j$, $\|x_i - x_j\| \to 0$ when $t \to +\infty$.*

The basic idea of a consensus protocol is that each individual agent makes use of the communication network to exchange information and accordingly designs an appropriate distributed control algorithm such that the complicated system resulting from the combination of agents' dynamics and network topology reaches consensus or synchronization. If the interaction network is capable of continuous communication, or if the network bandwidth is sufficiently large, the updating of the state information of each individual agent can be modeled by a differential equation. On the other hand, if the information exchange is performed in discrete fashion, the difference equation is then utilized to illustrate the updating of the agents' state information.

As is well known, in the past few decades, stochastic systems have gained more and more research interest and found wide applications in many fields because the stochastic phenomenon is inevitable and cannot be avoided in real-world systems. Therefore, the analysis and synthesis problems concerning stochastic systems have attracted much attention; see [105, 147, 176, 180–182] and the references therein. In the context of MASs, the randomness might occur in many different ways. First, it could happen in the system dynamics. For example, the random noises stemming from either an internal device or the external environment will impose stochasticity in both process and measurement equations. Moreover, an abrupt change of working condition, or the aging or erosion of certain equipment will also bring randomness in the system dynamics. Second, random switching of the network topology, which is frequently seen in practical engineering, would also make the MAS a stochastic system. Third, utilization of the communication network to transmit information will definitely face many issues such as time delay, which, as a matter of fact, will probably take place in a random way. Furthermore, the topology according to which the agents share their information might be varying stochastically.

All those aforementioned stochastic sources will bring about difficulties in both analysis and synthesis of the consensus problem for multi-agent systems. This is mainly because the traditional consensus definition that is proposed for deterministic dynamics and requires all the states of agents to converge to the same point will be no longer applicable in the context of stochastic MASs due to the existence of randomness. As a consequence, further development on consensus control for stochastic multi-agent systems is a major concern for practitioners and scientists from the areas of systems sciences and control engineering.

In the following, we shall give a brief survey to provide a thorough enumeration, classification and analysis of recent contributions in the research on consensus control of stochastic multi-agent systems. Although there exist previous reviews that have investigated the progress on this topic for deterministic MASs, this work is a first deep overview of recent advances in branches of stochastic MASs, additionally outlining some vital future challenges that need to be addressed to ensure meaningful progress and development of novel methods.

1.7   Consensus of Nonlinear Multi-Agent Systems with Stochastic Dynamics

In this section, we shall first give an overview of consensus control for multi-agent systems whose dynamics are governed by a stochastic differential equation or stochastic difference equation. Then, we shall review the consensus control problem in which the communication channels are affected by random noises, which make the MASs under investigation stochastic systems.

Due to the existence of stochasticity, the conventional consensus definitions for deterministic MASs, which require that all the agents' states converge to the same point, are not capable of characterizing the consensus process in stochastic contexts. Therefore, by borrowing ideas from stochastic analysis and control theory, researchers as well as engineers have put forward several consensus conceptions in different probabilistic senses to describe the consensus behaviors of stochastic MASs. These conceptions include but are not limited to mean square consensus, consensus in p-th moment, consensus in probability and almost sure consensus. In the following, we shall review the recent progress from the perspective of different locations where the stochasticity occurs in the system dynamics.

Stochasticity in process equations of agents

Different from most of the multi-agent consensus literature dealing with L_2 energy-bounded noises [1, 148, 160] or unknown-but-bounded noises, in this subsection, we shall review the recent progress on the consensus control problem for MASs whose process equations contain random noises. We will mainly discuss the problem from three aspects, i.e., consensus subject to additive noises, consensus subject to multiplicative noises (also known as state-dependent noises) and consensus subject to Markovian jump parameters.

Additive Noises

Additive noises are frequently seen in real-world engineering practice such as maneuvering target tracking [54], distributed sensing [20,35], flight control [17,21], and UAV [75,261], to name but a few. Working in a noisy environment, the dynamics of MASs can naturally be modeled by differential/difference equations subject to additive stochastic noises. The multi-agent system with additive noises in the process equations can be represented by the following state-space equation:

$$\dot{x}_i(t) = Ax_i(t) + Bu_i(t) + Dw_i(t) \tag{1.1}$$

where $x_i(t)$ and $u_i(t)$, respectively, stand for the state and control input of the i-th agent; $w_i(t)$ represents the additive process noises satisfying certain statistical properties and all the matrices are with compatible dimensions.

Based on the model (1.1), some representative work can be summarized as follows. The optimal control problem has been discussed in [209] for continuous MASs in which each of the sub-systems is subject to additive stochastic noises. The objective is to drive the individual agent to reach certain predetermined target states. Due to the existence of stochasticity in the system dynamics, an optimal control scheme has been developed aiming at minimizing, in the mean square sense, the so-called accumulated joint cost. It is worth mentioning that the developed optimization control technique could be applied in the optimal consensus control of multi-agent systems by certain appropriate manipulations. On the other hand, in [252], the distributed tracking-type games have been taken into account for a class of coupled discrete MASs subject to additive random disturbances. By resorting to the Nash certainty equivalence principle, the authors have designed the controller guaranteeing that the states of all the agents converge to a certain required function of the population state average. In [265], due to the existence of Gaussian white noise in the state feedback input, the definition *consensus with probability one* has been proposed and utilized to deal with the finite-time consensus control problem. By using probability theory, the consensus problem has been solved in the framework of Lyapunov theory.

Multiplicative Noises

In much of practical engineering, apart from the additive noises, the MASs are always confronted with multiplicative noises (also known as state-dependent noises), which can be expressed by the following state-space equation:

$$x_i(k+1) = Ax_i(k) + Bu_i(k) + Mx_i(k)w_i(k) \qquad (1.2)$$

where $w_i(k)$ denotes the stochastic multiplicative noises and all the matrices are with appropriate dimensions.

For discrete-time MASs described by (1.2), in [42], with the purpose of dealing with the multiplicative noises that enter into both process and measurement equations of agents' dynamics, the authors have proposed a novel type of consensus, i.e., *consensus in probability*. Then, by borrowing the ideas from those publications concerning input-to-state stability, sufficient conditions have been established for the addressed stochastic MASs to reach the desired consensus in probability. Note that different from [265] where the *consensus with probability one* is discussed, the control strategy proposed in [40] has the ability to drive all the agents to reach consensus with any predetermined probability.

As far as the continuous case of MASs subject to multiplicative noises is concerned, the Itô-type stochastic systems have been playing a paramount role; these dynamics are shown as follows:

$$dx_i(t) = (Ax_i(t) + Bu_i(t))dt + (Ex_i(t) + Hu_i(t))dW(t) \qquad (1.3)$$

where $W(t)$ is a scalar Brownian motion.

Recently, in [198], the exponential consensus problem has been considered for a type of nonlinear multi-agent system whose dynamics are governed by Itô-type differential equations. Sufficient conditions have been derived for the existence of the desired control protocol by means of the Lyapunov theory and comparison principle, which is capable of guaranteeing exponential leader-following consensus in the mean square sense. Likewise, a linear matrix inequality (LMI) based framework has been established in [58] for a class of Itô stochastic multi-agent systems to solve the passivity-based mean square exponential consensus control problem.

By virtue of Lyapunov theory, for a second-order stochastic multi-agent system which is governed by nonlinear Itô stochastic differential equations, control protocol has been designed in [264] with the purpose of driving all the agents to reach finite-time consensus in probability. Within a similar theoretical framework, the containment control in the mean square and mean square leader-following consensus problems have been discussed in [114] and [247], respectively, by resorting to Lyapunov theory in combination with the LMI approach.

Lately, in [227], exponential consensus in the mean square sense has been proposed to characterize the consensus process for a class of MASs with asynchronous switching topology. In this work, the time delays that stem from receiving confirmation of the mode switching before applying the matched controller have also been taken into account. By means of the so-called extended comparison principle, several easy-to-verify conditions for the existence of an asynchronously switched distributed controller are derived such that the addressed stochastic delayed multi-agent systems with asynchronous switching and nonlinear dynamics can achieve global exponential consensus in the mean square. Furthermore, in the technical note [228], the authors have investigated the stability problem of stochastic delayed systems by using a more general Halanay-type inequality with time-varying coefficients. A new comparison principle of the proposed Halanay-type inequality is developed to ensure the stability of stochastic delayed systems. The p-th moment stability, p-th moment asymptotic stability and p-th moment exponential stability are investigated by means of the proposed general Halanay-type inequality.

Markovian Jump Dynamics

In recent years, the Markovian jump system (MJS) has attracted tremendous research interest within the systems science and control communities owing primarily to its ability in modeling random variations; see [48, 49, 98–100, 111, 126, 137, 202] and the references therein. Accordingly, the corresponding analysis and design problems for multi-agent systems subject to Markovian parameters have started to gain ever-increasing attention. Some recent representative results and the corresponding developed approaches can be summarized as follows.

In [215], an output feedback control scheme has been presented for the stochastic MASs whose dynamics equations as well as performance index are expressed in terms of Markovian jump parameters. By virtue of the Markov jump optimal filtering theory in combination with the mean field approach, the distributed feedback control algorithm has been designed to drive all the agents to approach the desired position. For heterogeneous multi-agent systems, in [268], a stochastic model has been introduced to characterize the switching between two types of agents, that is, active agents and passive agents. The heterogeneous structure of the addressed MAS has been cast into an appropriate chosen Markov chain and a unified framework has then been established within which a necessary and sufficient condition has been given to ensure the requested consensus. In [1], the synchronization problem has been first solved for a class of complex dynamical networks with Markovian jump parameters that are switching among a finite set. Then, the obtained results have been extended to the relevant multi-agent system with stochastic time delays. Broadcast control has been investigated in [8] for a group of Markovian multi-agent systems. A sufficient condition has been presented for the addressed Markovian MAS to stabilize the resulting broadcast control systems.

On the other hand, in addition to the utilization of describing system dynamics, the Markov chain has also been extensively employed to characterize quite a few other frequently seen stochastic phenomena in consensus control problems for MASs. For instance, the data missing phenomenon which usually occurs in information transmitting among agents has been handled in [159], where the Markov chain has been adopted to depict the random nature of the data missing. Likewise, the Markov chain has been employed in [80] to characterize the random input time delay in a class of multi-agent systems.

By resorting to the Lyapunov theory and linear matrix inequality approach, necessary and sufficient conditions have been derived to achieve the desired consensus.

Recently, with regard to second-order MASs, impulsive control methodology has been exploited in [242] where sampled information used for generating the feedback control input is subject to random heterogeneous time delays governed by a Markov chain. Similarly, in [241], two mutually independent Markov chains have been employed to characterize the stochastic switching topology and random communication delays. The key idea behind the proposed approach is to convert the original systems into an expanded analogous error system with two Markovian parameters by means of model transformation.

Within the $\mathcal{H}_\infty$ framework, the control routing strategy has been proposed in [1] for mobile multi-agent networked systems. The Markovian jump linear system theory has been used to develop the required decentralized scheme with a prescribed $\mathcal{H}_\infty$ disturbance attenuation level. Such an $\mathcal{H}_\infty$ framework has also found applications in [148], where fault detection and isolation problems have been dealt with for a class of discrete-time MASs with Markovian jump parameters. In this work, the imperfect communication channel and stochastic packet dropping effects have also been taken into consideration. Sufficient conditions have been established where the required fault detection and isolation algorithms have been proposed with an $\mathcal{H}_\infty$ prescribed performance criterion.

Other Random Phenomena

Apart from the aforementioned literature concerning the consensus control problem subject to stochastic noises, random data missing and random communication time delays, some literature has been concerned with corresponding problems with other random phenomena. For instance, stochastic communication failure occurring in the data transmission channels has been considered in [86], whereas the randomly changing leader following issue has been taken into consideration in [174] where the leader is changing stochastically.

Stochasticity in output/measurement equations of agents

Consider the following multi-agent systems:

$$\begin{cases} \dot{x}_i(t) = Ax_i(t) + Bu_i(t) \\ y_i(t) = Cx_i(t) + Dv_i(t) \\ u_i(t) = -\sum_{j \in \mathcal{N}_i(t)} a_{ij}(t)(y_j(t) - y_i(t)) \end{cases} \quad (1.4)$$

where $y_i(t)$ stands for the measurement output of agent i; $v_i(t)$ denotes the additive random noises corrupting the measurements; the real-valued coefficient $a_{ij}(t)$ represents the weighting parameter imposed on the information sent to agent i from agent j. Here, if at time t, agent j has not sent information to agent i, then $a_{ij}(t) = 0$. All the matrices are of compatible dimensions. From (1.4), we can clearly see that the stochasticity appearing in the measurement output will impact the consensus performance via propagation of information among the agents according to the communication topologies.

For discrete-time MAS with additive noises in measurements, several consensus notions in the probabilistic sense, such as *weak consensus, strong consensus, mean square (m.s.) consensus* and *almost sure (a.s.) consensus* have been proposed in [88, 96, 113] to characterize the consensus behavior of the addressed stochastic multi-agent systems. A unified framework has been established within which the proposed consensus problems in the stochastic context can be discussed. By resorting to the developed *stochastic approximation-type algorithms* using a decay factor with adjustable decreasing step size, all the individual states can be driven towards each other, thereby achieving the required consensus in the desired stochastic sense.

Similar notions have been extended in [118, 120] for continuous stochastic MASs, where the definitions of *average-consensus in the p-th moment* and *almost sure (a.s.) consensus* have been proposed to cope with the stochastic consensus problem for a class of multi-agent systems whose measurements are affected by Gaussian white noises. In [112], a necessary and sufficient condition has been derived for the existence of an asymptotic unbiased protocol, ensuring the *mean square average-consensus* of the considered MASs with measurement noises.

With additive measurement noises, the work in [269] has been concerned with stochastic consensus of multi-agent systems where the information exchange among agents is described by a directed graph. By combining algebraic graph theory, matrix theory and stochastic analysis, stochastic weak and strong consensus are examined. For the case with general digraphs, the authors have given the necessary and sufficient conditions for the almost sure strong consensus and show that the mean square strong consensus and almost sure strong consensus are equivalent. Moreover, some necessary conditions and sufficient conditions have been obtained for the mean square and almost sure weak consensus. Especially, the necessary and sufficient conditions for the mean square weak consensus are given for the case with undirected graphs.

For discrete MASs, in [3], the mean square consensus has been replaced by the so-called approximated consensus because of the adoption of local voting protocols with non-vanishing (e.g., constant) step size. Such an approximation methodology for consensus in a noisy environment was originated in [88, 96], where it was proven that if the random noises have certain statistical properties (i.e., zero mean, bounded variances), then consensus can always be achieved in the almost sure sense by means of adopting a monotonically decreasing gain algorithm. The key of the proposed consensus schemes is to solve the nonlinear equations with measurements corrupted by noises thereby conducting recursive stochastic approximations procedures [30, 108, 179, 195].

Recently, with the help of the aforementioned stochastic approximation technique, a similar idea has been extended in [158] to cope with consensus issues subject to noisy environments by resorting to averaging approaches with stopping rules. It should be pointed out that the number of iterations in the stochastic approximation procedure has been given explicitly in virtue of the coefficients relevant to the desired consensus performance (accuracy or precision) as well as the probabilistic constraints. The similar framework has been established in [157] for a class of MASs with noisy communication, where a stopping rule oriented stochastic approximation method has been given in terms of characteristic values of communication network graphs. In comparison to the aforementioned literature concerning steady state performance of the multi-agent systems, this paper has made the first attempt to investigate transient performance during the consensus process.

The aforementioned literature is concerned with the case of additive noises. It should be mentioned here that in [270], the corresponding consensus issue has been discussed while the measurement outputs are affected by multiplicative measurement noises. By virtue of algebraic graph theory and matrix theory, the consensus problem has been converted into the stochastic stability problem of stochastic differential equations (SDEs) driven by multiplicative noises. In this work, both almost sure (a.s.) consensus and mean square (m.s.) consensus have been taken into account and the corresponding sufficient condition has been established. It is shown that, for any bounded noise intensities, the desired mean square and the almost sure consensus can be achieved by carefully choosing the control gain. In [119], the conception of average consensus in p-th has been firstly proposed and then applied to solve the consensus problem for a class of continuous MAS subject to communication noises that are described as multiplicative noises dependent on the states in the output measurements. Based on Lyapunov theory and the stochastic analysis technique, sufficient conditions for the addressed consensus problem have been established.

Stochasticity in communication channels

In real-world applications, it is always the case that measurement output information is incomplete due to various reasons, such as random communication delays, random network topologies, stochastic Gaussian fading channels, which induce certain stochasticity in the analysis and synthesis issues in the consensus problem for multi-agent systems. The MASs subject to stochastic disturbances in communication channels are expressed as follows:

$$\begin{cases} x_i(k+1) = Ax_i(k) + Bu_i(k) \\ y_i(k) = Cx_i(k) \\ u_i(k) = -\sum_{j \in \mathcal{N}_i(k)} a_{ij}(k)(y_j(k) - y_i(k) + \delta_i(k)) \end{cases} \quad (1.5)$$

where $\delta_i(k)$ describes the noises or disturbances during the information propagation via communication networks.

Some recent representative works can be summarized as follows. For instance, a distributed consensus-based protocol named Average TimeSync has been derived in [184] where the case of random communication has been taken into account. In [226], the phenomenon of random time delay occurring in the communication channels has been modeled by introducing a Markov chain. By taking into account the transition probability of the time-delay, the authors have obtained a less conservative design algorithm for the consensus of the addressed MASs. For heterogeneous MASs, a similar problem has been solved in [243] by using an event-triggering tracking control algorithm. Very recently, in [203], the authors have studied a quite general synchronization problem while the communication networks are subject to random delays. In this work, the communication delays are assumed to be bounded and obeying some independent identical distribution with certain known statistical properties. A much more general problem has been taken into consideration in [216] for distributed networked control systems where various communication constraints such as time delays, noises and link failures are incorporated in a unified framework.

Aside from random communication delays, the channels used for signal transmitting always face the corruption of external stochastic noises or disturbances. Recently, for a class of multi-agent systems with noisy communication, a stopping rule oriented stochastic approximation method has been given in [157] in terms of characteristic values of communication network graphs. In comparison to the aforementioned literature concerning the steady state of the MASs, this paper makes the first attempt to deal with transient performance during the consensus process. Moreover, in [96], two algorithms have been designed to handle the average consensus problems subject to communication noises. In order to cope with the so-called bias-variance dilemma (which indicates that running consensus for a longer period leads to more accurate consensus performance, but could probably result in larger variance), two consensus algorithms have been proposed to reach a satisfactory balance between the two essential performance indices by playing with the tradeoffs. In this paper, similar as those mentioned in [3, 30, 108, 157, 158, 179, 195], the stochastic approximation arguments as well as the Markov process theory have been utilized to derive the condition for the addressed MASs to reach almost sure consensus.

In [113], the average consensus control problem has been investigated for the first-order discrete time multi-agent system subject to uncertain communication environments. By virtue of probability limit theory as well as algebraic graph theory, the addressed problem has been solved despite the existence of stochastic communication noises. For a continuous case, the authors in [164] have proposed a stochastic version of the averaging approach originated in [162, 163] to solve the mean square leader-following consensus problem with additive communication noises.

Stochasticity in sampling

As is well known, in real-world application, the signal received by the sensor of each agent has to be sampled before it is transferred or utilized since in the context of wireless communication networks the information is usually sent digitally. This gives rise to control or filtering algorithms using sampling techniques. Accordingly, consensus problems for multi-agent systems where various sampling approaches are utilized have been attracting ever-increasing research interest recently; see, e.g. [29, 38, 71, 78, 102, 121, 191, 192, 229, 231, 246, 258, 260] for examples. It should be mentioned that, however, due to the limitation of devices, sometimes the signals might not be captured, which leads to the undesirable phenomenon where the sampled information is lost in probability. In this case, the randomness in the sampling will inevitably impact the multi-agent systems through the propagation of measurement output via communication networks.

Recently, such an issue has started to attract research attention. For example, in [214], the mean square node-to-node consensus tracking issue has been investigated under a stochastically sampling mechanism. The original problem has been converted to the control problem for systems with stochastic time-delay, and therefore, sufficient conditions have been derived by adopting routine techniques for analysis and synthesis for stochastic time-delay systems (e.g., the Lyapunov functional approach, LMI algorithm, stochastic analysis, etc.).

In [221], the mean square exponential consensus problem with an $\mathcal{H}_\infty$ criterion has been solved by means of the linear matrix inequality approach within the Lyapunov theory framework. It should be emphasized that, in this work, the sampling period of each agent switches between two possible values in a probabilistic way. Following a similar line, the stochastic sampled-data leader-following consensus problem of nonlinear MASs has been addressed in [78]. By introducing certain random variables obeying the Bernoulli distribution, such a stochastic sampling phenomenon is properly characterized. As for the continuous case, a similar problem has been considered in [83] where the sampling instants have been assumed to be stochastic and taking values in a finite set. Based on such a stochastic sampling mechanism, the event-triggering technique has also been employed by making use of the sampled data to solve the corresponding consensus control problem.

1.8 Consensus with Stochastic Topologies

In this section, we proceed to make a review of some recent results concerning the consensus problems for MASs with stochastic topologies.

So far, consensus control for a multi-agent system subject to stochastic topologies has mainly focused on Markovian switching communication topologies, see e.g. [190, 262]. This is primarily due to the ability of the Markov chain in modeling the variation in the random changes of the communication situations during the data transmission among agents. Some representative publications can be summarized as follows.

For a class of heterogeneous MASs comprised of first and second order agents, the mean square consensus control problem has been investigated in [155] where the topology switches in a finite set according to a known Markov chain. The necessary and sufficient condition has been established for the requested consensus by means of a stochastic irreducible aperiodic matrix method. Mean square containment control has been discussed in [149] under the same Markovian switching topology. The similar mean square consensus control problem has been solved in [150] for nonlinear continuous multi-agent systems.

In [175], the considered topology is assumed to be not only switching according to a known Markovian chain, but also subject to time-delays. The time-delays under investigation are entering the MAS dynamics from both measurement transmission and state feedback. In the case of the simultaneous occurrence of Markovian topologies, communication noises and delays, an ergodicity approach has been proposed in [152] for backward products of degenerating stochastic matrices via a discrete time dynamical system approach to analyze the consensus stabilization problem. The relevant problems have been solved in [43,153,225,251] where the addressed MASs are with Markovian switching topologies, communication noises and/or time-delays.

In addition to the mean square consensus, several other consensus definitions and performance indices have been proposed to characterize different consensus process/requirements for MASs subject to Markovian topologies. For instance, in [263], a necessary and sufficient condition has been derived for a class of MASs to achieve the expected consensus with the guarantee of minimum communication cost. An $L_2 - L_\infty$ performance criterion has been proposed in [160] for the studied MAS to reach the desired consensus with satisfactory steady state performance. Different from some existing results, global jumping modes of the Markovian network topologies have not been required to be completely available for consensus protocol design. By developing a so-called network topology mode regulator, the unavailable global modes have been decomposed into several overlapping groups. Then, a new group mode-dependent distributed consensus protocol on the basis of relative measurement outputs of neighboring agents has been proposed.

It is worth mentioning that, different from the aforementioned literature regarding Markovian switching topologies, in [31], a series of mutually independent Bernoulli sequences have been introduced to illustrate the random switching topology. The resulting multi-agent system is named Bernoulli network and the considered mean square consensus problem can be converted into the mean square stabilization problem of delayed systems subject to the stochastic switching signal. Sufficient conditions for the desired consensus performance have then been established within an LMI framework.

1.9 Outline

The outline of this book is given as follows. It it worth mentioning that Chapter 2–Chapter 4 are mainly focused on the sliding mode control problems for networked time-invariant systems, while in Chapter 5 concerns the networked time-varying system. Chapter 6–Chapter 8 investigate consensus control problems for time-varying multi-agent systems in different scenarios, whereas Chapter 9–Chapter 10 discuss the distributed filtering problems over sensor networks. The framework of this book is shown as follows:

- In Chapter 1, research background is first introduced, which mainly involves the nonlinear stochastic networked systems, sliding mode control, stochastic time-varying multi-agent systems, randomly occurring network-induced phenomena, sensor networks and distributed filtering. Then the outline of the book is listed.

- In Chapter 2, a robust SMC design problem for a class of uncertain nonlinear stochastic systems with multiple data packet losses is studied. Both matched and unmatched nonlinearities are taken into consideration. The multiple data packet losses are assumed to happen in a random way, and the loss probability of each individual state variable is governed by a corresponding individual stochastic variable obeying a certain probabilistic distribution in the interval $[0 \ 1]$. We also introduce, for the first time, a stochastic switching function for the SMC problem of discrete-time stochastic systems. By means of LMI, a sufficient condition for the exponentially mean square stability as well as pre-specified $\mathcal{H}_\infty$ performance index of the system dynamics on the specified sliding surface is derived. By the reaching condition proposed in this chapter, an SMC controller is designed to globally drive the state trajectory onto the specified surface with probability 1, which gives rise to a non-increasing zigzag motion along the surface.

- Chapter 3 investigates the SMC design problem for a class of uncertain nonlinear networked systems with multiple stochastic communication delays. By means of LMIs, a sufficient condition for the robustly exponential stability of the system dynamics on the specified sliding surface is derived. By the reaching condition proposed in this chapter, an SMC controller is designed to globally drive the state trajectory onto the specified surface, which gives rise to a non-increasing zigzag motion along the surface.

- Chapter 4 is concerned with the SMC control problem for a type of uncertain nonlinear networked systems with multiple stochastic communication delays. A new form of discrete switching function is put forward for the first time in this chapter. By means of LMIs, a sufficient condition for the globally mean square asymptotic stability of the system on the specified sliding surface is derived. By the reaching condition applied in this chapter, an SMC controller is designed that globally drives the state trajectories onto the specified surface with probability 1.

- In Chapter 5, we deal with the reliable $\mathcal{H}_\infty$ control problem for a type of nonlinear stochastic system with sensor failures. The stochastic nonlinearities taken into consideration could cover several well-studied nonlinearities. The sensor failures occur in a random way, and the failure probability is described by a stochastic variable satisfying the Bernoulli distribution. The solvability of the addressed control problem is expressed by the feasibility of certain linear matrix inequalities. The numerical values of the controller gains can be obtained by the given computing algorithm.

- Chapter 6 is concerned with the event-triggered mean square consensus control problem for a class of discrete time-varying stochastic multi-agent systems subject to sensor saturations. First, a new definition of mean square consensus is presented for the addressed MAS to characterize the transient consensus behavior. Then, by means of an RLMI approach, sufficient conditions are established for the existence of the desired controller. Within the established framework, two optimization problems are discussed to optimize the consensus performance and triggering frequency, respectively.

- In Chapter 7, the mean square $\mathcal{H}_\infty$ consensus control problem is investigated for a class of nonlinear discrete time-varying stochastic multi-agent systems. The stochastic nonlinearities characterized by statistical means are quite general and could cover several well-studied nonlinearities as special cases. By means of the recursive linear matrix inequality approach, a general framework is established for the addressed multi-agent system to reach the desired consensus satisfying both $\mathcal{H}_\infty$ specification and mean square criterion. Sufficient conditions are derived in terms of a set of recursive matrix inequality.

- In Chapter 8, the quasi-consensus control problem is investigated for a class of discrete time-varying nonlinear stochastic multi-agent systems subject to deception attacks. A new definition of quasi-consensus is first presented to characterize the consensus process where all the agents are constrained to stay within a certain ellipsoidal region at each time step. A novel deception attack model is proposed, where the attack signals are injected by the adversary into the measurement data during the process of information transmission via the communication network. By resorting to the recursive matrix inequality approach, sufficient conditions are established for the solvability of the quasi-consensus control problem. Subsequently, an optimization problem is provided to determine the feedback gain parameters that guarantee the locally optimal consensus performance.

- In Chapter 9, we are concerned with the distributed event-based set-membership filtering problem for a class of discrete nonlinear time-varying systems subject to unknown but bounded noises and sensor saturations over sensor networks. A novel event-triggering communication mechanism is proposed for the sake of reducing the sensor data transmission rate and energy consumption. By means of the recursive linear matrix inequalities approach, the sufficient conditions are established for the existence of the desired distributed event-triggering filter. With the established framework, two optimization problems are discussed to demonstrate the flexibility of the proposed methodology in making trade-offs between accuracy and cost. Finally, a numerical simulation example is exploited to verify the effectiveness of the distributed event-triggering filtering strategy.

- Chapter 10 investigates the variance-constrained distributed filtering problem for a class of discrete time-varying systems with multiplicative noises, unknown but bounded disturbances and deception attacks over sensor networks. A novel deception attack model is proposed, where the attack signals are injected by the adversary into both control and measurement data during the transmission via the communication networks. A sufficient condition is established for the existence of the required filter satisfying the estimation error variance constraints by means of the RLMI approach. An optimization problem is presented to seek the filter parameters with the guarantee of the locally minimal estimation error variance at each time instant.

- In Chapter 11, the conclusions and some potential topics for future work are given.

2

Robust $\mathcal{H}_\infty$ Sliding Mode Control for Nonlinear Stochastic Systems with Multiple Data Packet Losses

In the past few decades, sliding mode control (also known as variable structure control) has been extensively studied because of its advantage of strong robustness against model uncertainties, parameter variations and external disturbances. In sliding mode control, trajectories are forced to reach a sliding manifold in finite time and then stay on the manifold for all future time. It is worth mentioning that, in the existing literature concerning the sliding mode control (SMC) problem for nonlinear systems, the nonlinearities and uncertainties taken into consideration are mainly under matching conditions; that is to say, the nonlinear and uncertain terms enter the state equation at the same point as the control input and consequently the motion on the sliding manifold is independent of those matched terms. However, in engineering practice, a large part of external nonlinear disturbances and parameter uncertainties cannot be treated as a matched type of nonlinearities. In recent years, since most control strategies are implemented in a discrete-time setting (e.g., networked control systems), the SMC problem for discrete-time systems has gained considerable research interest and many results have been reported in the literature.

On another research front, in most practical systems nowadays such as a target tracking system, there may be certain observations that consist of noise only when the target is absent due to its high maneuverability. In other words, the measurements are not consecutive but usually subject to partial or complete information missing. Such a phenomenon is referred to as measurement missing, information dropout or data packet losses, which occurs frequently for a variety of reasons such as sensor temporal failure, network congestion, accidental loss of some collected data or network-induced delay, and might lead to system performance degradation and sometimes even instability. Therefore, in the past few years, a great deal of research effort has been made to solve the control and filtering problems in the presence of data packet losses. Such a data packet loss phenomenon is usually characterized in a probabilistic way. Up to now, the $\mathcal{H}_\infty$ sliding mode control problem has not been studied for the discrete-time uncertain nonlinear stochastic system with multiple data packet losses. By using the discrete-time sliding motion concept, this chapter aims to design a state feedback controller such that 1) the system state trajectories are globally driven onto the pre-specified sliding surface with probability 1, resulting in a non-increasing zigzag motion on the sliding surface; 2) the exponentially mean-square stability and the $\mathcal{H}_\infty$ noise attenuation level of the system are simultaneously achieved on the pre-specified sliding surface.

The rest of this chapter is arranged as follows. Section 2.1 formulates an uncertain nonlinear stochastic system with multiple data packet losses. In Section 2.2, a novel switching function is first put forward and then two LMI-based sufficient conditions are given to obtain the parameters in the proposed switching function for simultaneously ensuring the exponentially mean-square stability and $\mathcal{H}_\infty$ performance on the sliding surface. Secondly, an SMC law is synthesized to drive the state trajectories onto the specified surface with probability 1. In Section 2.3, an illustrative numerical example is provided to show the effectiveness and usefulness of the proposed approach. Section 2.4 gives our conclusions.

2.1   Problem Formulation

Consider an Itô-type nonlinear stochastic system governed by the following state-space equation:

$$x(k+1) = (A+\Delta A)x(k)+B(u(k)+f(x(k)))+E_1g(x(k))+E_2x(k)\omega(k) \quad (2.1)$$

and the output equation:

$$y(k) = Cx(k) + E_3h(x(k)) + D\nu(k) \quad (2.2)$$

where $x(k) \in \mathbb{R}^n$ is the state vector, $y(k) \in \mathbb{R}^p$ is the output signal, $u(k) \in \mathbb{R}^m$ is the control input, $\nu(k) \in l_2$ is a stochastic external disturbance. A, B, C, D, E_1, E_2 and E_3 are known constant real-valued matrices with appropriate dimensions. The nonlinear function $f(x(k))$ represents the matched bounded disturbance. $\omega(k)$ is a scalar Wiener process (Brownian Motion) on $(\Omega, \mathcal{F}, \mathcal{P})$ with

$$\mathbb{E}\left\{\omega(k)\right\} = 0,$$
$$\mathbb{E}\left\{\omega^2(k)\right\} = 1, \quad (2.3)$$
$$\mathbb{E}\left\{\omega(k)\omega(j)\right\} = 0 \quad (k \neq j).$$

The matrix ΔA is the real-valued norm-bounded parameter uncertainty

$$\Delta A = MFN \quad (2.4)$$

where M and N are known real constant matrices which characterize how the deterministic uncertain parameter in F enters the nominal matrix A with

$$F^{\mathrm{T}}F \leq I. \quad (2.5)$$

The parameter uncertainty ΔA is said to be admissible if both (2.4) and (2.5) are satisfied.

The vector-valued nonlinear functions $g(x(k))$ and $h(x(k))$ stand for the unmatched external nonlinearities, satisfying:

$$
\begin{aligned}
&[g(x) - g(z) - U_1(x - z)]^{\mathrm{T}} [g(x) - g(z) - U_2(x - z)] \le 0, \\
&\qquad\qquad g(0) = 0, \quad \forall x, z \in \mathbb{R}^n, \\
&[h(x) - h(z) - V_1(x - z)]^{\mathrm{T}} [h(x) - h(z) - V_2(x - z)] \le 0, \\
&\qquad\qquad h(0) = 0, \quad \forall x, z \in \mathbb{R}^n
\end{aligned}
\tag{2.6}
$$

where U_1, U_2, V_1 and $V_2 \in \mathbb{R}^{n \times n}$ are known real constant matrices, with $U = U_1 - U_2$ and $V = V_1 - V_2$ being positive definite matrices.

In this chapter, the phenomenon of multiple data packet losses, which frequently occurs in a networked environment, is also taken into consideration. We use the following formula to describe such a multiple data packet losses situation:

$$
\bar{x}(k) = \Theta x(k)
\tag{2.7}
$$

where $\bar{x}(k)$ is the actual signal obtained from the process of sampling the feedback or output signal. The matrix Θ is defined as

$$
\Theta = \mathrm{diag}\{\theta_1(k), \theta_2(k), \cdots, \theta_n(k)\}
$$

with $\theta_i(k)$ $(i = 1, 2, \cdots, n)$ being n unrelated random variables which are also unrelated with $\omega(k)$. It is assumed that $\theta_i(k)$ has the probabilistic density function $\varrho_i(s)$ $(i = 1, 2, \cdots, n)$ on the interval $[0\ \ 1]$ with mathematical expectation μ_i and variance σ_i^2. Note that $\theta_i(k)$ could satisfy any discrete probabilistic distributions on the interval $[0\ \ 1]$. Due to multiple data packet losses (2.7), the output equation (2.2) should be amended as

$$
\begin{aligned}
y(k) &= C\bar{x}(k) + E_3 h(x(k)) + D\nu(k) \\
&= C\Theta x(k) + E_3 h(x(k)) + D\nu(k) \\
&= \sum_{i=1}^{n} C_i \theta_i(k) x(k) + E_3 h(x(k)) + D\nu(k)
\end{aligned}
\tag{2.8}
$$

where

$$
C_i \triangleq C \cdot \mathrm{diag}\{\underbrace{0, \cdots, 0}_{i-1}, 1, \underbrace{0, \cdots, 0}_{n-i}\}.
\tag{2.9}
$$

In the sequel, we denote $\bar{\Theta} = \mathbb{E}\{\Theta\}$.

Remark 2.1 *In the formulation of the multiple data packet losses (2.7), $\theta_i(k)$ could take value on the interval $[0\ \ 1]$; hence it includes the widely used Bernoulli distribution as a special case. To be specific, when $\theta_i(k) = 0$ (respectively, $0 < \theta_i(k) < 1$), the ith state variable $x_i(k)$ is completely (respectively, partially) lost at the sampling instant k. The main difference between* *the model for output missing proposed in [222, 238] and this chapter is that the former focuses on the data missing phenomenon caused by sensors' failures while, in this chapter, we are interested in the situation that the data packet, due to complex circumstances such as network congestion and transmission lines aging, is completely or partially lost before it reaches the sensors.*

2.2   Design of Sliding Model Controllers

In this section, we first propose a switching function in a *stochastic* form for the uncertain nonlinear system (2.1) with data packet losses. Then, two theorems will be given in order to design the switching function parameters capable of simultaneously ensuring the exponentially mean square stability and the $\mathcal{H}_\infty$ performance in the sliding motion. It is shown that the controller design problem in the sliding motion can be solved if an LMI with an equality constraint is feasible. Finally, a controller is synthesized to satisfy the improved discrete-time sliding motion reaching the condition to drive the trajectories of system (2.1) onto the pre-specified sliding surface with probability 1.

Sliding Surface

In this chapter, considering the existence of random data packet losses in the feedback loop, we choose the switching function as follows:

$$s(k) = G\bar{x}(k) = G\Theta x(k) \tag{2.10}$$

where G is designed such that $G\bar{\Theta}B$ is nonsingular and $G\bar{\Theta}E = 0$, where $E \triangleq \begin{bmatrix} E_1 & E_2 \end{bmatrix}$. In this chapter, we select $G\bar{\Theta} = B^\mathrm{T}P$ with $P > 0$ being a positive definite matrix to confirm the non-singularity of $G\bar{\Theta}B$.

It can be seen that the switching function (2.10) serves as a stochastic difference equation due to the existence of the random variable matrix Θ. Therefore, the traditional necessary condition for discrete-time quasi-sliding motion, stated as $s(k+1) = s(k) = 0$, should be re-formulated on $(\Omega, \mathcal{F}, \mathcal{P})$ as follows:

$$\mathcal{P}\{s(k+1) = s(k) = 0\} = 1. \tag{2.11}$$

In order to obtain the equivalent control law of the sliding motion, we take

$$\mathbb{E}\{s(k+1)\} = \mathbb{E}\{s(k)\} = 0. \tag{2.12}$$

Solving the above for $u(k)$, the equivalent control law of the sliding motion is given by

$$u_{eq}(k) = -(B^\mathrm{T}PB)^{-1}B^\mathrm{T}P(A + \Delta A)x(k) - f(x(k)). \tag{2.13}$$

Substituting (2.13) as $u(k)$ into (2.1) yields

$$x(k+1) = A_K x(k) + E_1 g(x(k), k) + E_2 x(k) \omega(k) \qquad (2.14)$$

where $A_K \triangleq A + \Delta A - B(B^{\mathrm{T}} P B)^{-1} B^{\mathrm{T}} P(A + \Delta A)$. The expression (2.14) is the sliding mode dynamics of system (2.1) in the specified switching surface $\mathbb{E}\{s(k+1)\} = \mathbb{E}\{s(k)\} = 0$.

Remark 2.2 *It is the first time in the literature that a stochastic switching function (2.10) is introduced to deal with the discrete-time SMC problem for stochastic systems with multiple random data packet losses. The reason why we use (2.12) as the necessary condition for a discrete-time sliding motion is that it is meaningless to solve a stochastic difference equation $s(k+1) = s(k) = 0$ for an equivalent control law in sliding motion followed by the deterministic SMC controller parameters. As a result, both the equivalent control law (2.13) and the sliding mode dynamics (2.14) exist in a probabilistic sense. Moreover, throughout this chapter, the SMC controller is designed to satisfy the discrete-time sliding motion reaching condition as well as the necessary condition on the sliding surface also in a probabilistic sense.*

Remark 2.3 *In the output equation (2.8) and stochastic switching function (2.10), the matrix Θ is employed to describe the random data packet losses in the output channel as well as in the state feedback loop. Generally, in most real-world engineering practices, the probabilities of data packet losses through feedback channel and output channel might not be identical to each other since they are always transmitted by different ways. Nevertheless, the packet loss probabilities are assumed to be the same in this chapter purely to avoid unnecessarily complicated notations. It should be pointed out that our main results can be easily extended to more general cases where different data transmitting channels have different packet loss probabilities.*

Remark 2.4 *The condition $G\bar{\Theta}E = 0$ is applied to eliminate the unmatched nonlinearity $g(x(k))$ and the Brownian motion on the sliding surface so as to obtain the deterministic form of switching parameters that will be used to synthesize the SMC controller. For continuous-time stochastic systems, such a methodology has been used in [166, 167], where the unmatched external nonlinearity is not taken into consideration. It is worth mentioning that, by the proposed technique in this chapter, we could deal with a wide range of nonlinearities, either stochastic or deterministic.*

Before stating the designing goal, we introduce the following stability concept for system (2.14).

Definition 2.1 *The system (2.14) is said to be robustly mean square stable if, for any $\varepsilon > 0$, there exists a $\delta(\varepsilon) > 0$ such that $\mathbb{E}\{\|x(k)\|^2\} < \varepsilon$ $(k > 0)$ when $\mathbb{E}\{\|x(0)\|^2\} < \delta(\varepsilon)$. And if $\lim_{k \to \infty} \mathbb{E}\{\|x(k)\|^2\} = 0$ for any $x(0) \in \mathbb{R}^n$, then the system (2.14) is said to be asymptotically mean square stable.*

Moreover, if there exist constants $\beta \geq 1$ and $0 < \tau < 1$ such that $\mathbb{E}\{\|x(k)\|^2\} \leq \beta\tau^k\mathbb{E}\{\|x(0)\|^2\}$, then the system (2.14) is said to be exponentially mean square stable.

In this chapter, we aim to synthesize an SMC law such that, for all admissible parameter uncertainties and multiple data packet losses, the following two requirements are achieved simultaneously:

(Q1) The state trajectory of system (2.1) is globally driven onto the pre-specified sliding surface (2.10) with probability 1 and, subsequently, the sliding motion is exponentially mean square stable.

(Q2) For a given scalar $\gamma > 0$, with $\nu(k) \neq 0$, the controlled output $y(k)$ satisfies

$$\sum_{k=0}^{\infty} \mathbb{E}\left\{\|y(k)\|^2\right\} \leq \gamma^2 \sum_{k=0}^{\infty} \mathbb{E}\left\{\|\nu(k)\|^2\right\}, \tag{2.15}$$

under the zero initial condition.

The problem addressed above is referred to as the robust $\mathcal{H}_\infty$ sliding mode control for nonlinear stochastic systems with multiple packet losses.

Stability and $\mathcal{H}_\infty$ Performance on a Sliding Surface

In this subsection, we present two theorems to determine the parameters appearing in switching function (2.10). These parameters are necessary for designing the SMC controller to fulfill the control tasks (*Q1*) and (*Q2*).

To begin with, we introduce the following lemmas which will be used later.

Lemma 2.1 *Let $\mathcal{W}(k) = x^{\mathrm{T}}(k)Px(k)$ be a Lyapunov functional where $P > 0$. If there exist real scalars ζ, μ, υ and $0 < \psi < 1$ such that both*

$$\mu\|x(k)\|^2 \leq \mathcal{W}(k) \leq \upsilon\|x(k)\|^2 \tag{2.16}$$

and

$$\mathbb{E}\{\mathcal{W}(k+1)|x(k)\} - \mathcal{W}(k) \leq \zeta - \psi\mathcal{W}(k) \tag{2.17}$$

hold, then the process $x(k)$ satisfies

$$\mathbb{E}\{\|x(k)\|^2\} \leq \frac{\upsilon}{\mu}\|x(0)\|^2(1-\psi)^k + \frac{\lambda}{\mu\psi}. \tag{2.18}$$

Lemma 2.2 *For any real vectors a, b and matrix $P > 0$ of compatible dimensions,*

$$a^{\mathrm{T}}b + b^{\mathrm{T}}a \leq a^{\mathrm{T}}Pa + b^{\mathrm{T}}P^{-1}b. \tag{2.19}$$

Lemma 2.3 *(Schur Complement Lemma) Given constant matrices $\mathcal{S}_1, \mathcal{S}_2, \mathcal{S}_3$ where $\mathcal{S}_1 = \mathcal{S}_1^{\mathrm{T}}$ and $0 < \mathcal{S}_2 = \mathcal{S}_2^{\mathrm{T}}$, then $\mathcal{S}_1 + \mathcal{S}_3^{\mathrm{T}}\mathcal{S}_2^{-1}\mathcal{S}_3 < 0$ if and only if*

$$\begin{bmatrix} \mathcal{S}_1 & \mathcal{S}_3^{\mathrm{T}} \\ \mathcal{S}_3 & -\mathcal{S}_2 \end{bmatrix} < 0 \quad or \quad \begin{bmatrix} -\mathcal{S}_2 & \mathcal{S}_3 \\ \mathcal{S}_3^{\mathrm{T}} & \mathcal{S}_1 \end{bmatrix} < 0. \tag{2.20}$$

Lemma 2.4 *(S-procedure) Let $J = J^\mathrm{T}$, M and N be real matrices of appropriate dimensions, and F satisfy (2.5). Then $J + MFN + N^\mathrm{T}F^\mathrm{T}M^\mathrm{T} < 0$ if and only if there exists a positive scalar ε such that $J + \varepsilon MM^\mathrm{T} + \varepsilon^{-1}N^\mathrm{T}N < 0$ or, equivalently,*

$$\begin{bmatrix} J & \varepsilon M & N^\mathrm{T} \\ \varepsilon M^\mathrm{T} & -\varepsilon I & 0 \\ N & 0 & -\varepsilon I \end{bmatrix} < 0. \tag{2.21}$$

Denote

$$\tilde{U} = \frac{U_1^\mathrm{T}U_2 + U_2^\mathrm{T}U_1}{2},$$

$$\bar{U} = \frac{-U_1^\mathrm{T} - U_2^\mathrm{T}}{2},$$

$$\tilde{V} = \frac{V_1^\mathrm{T}V_2 + V_2^\mathrm{T}V_1}{2},$$

$$\bar{V} = \frac{-V_1^\mathrm{T} - V_2^\mathrm{T}}{2}.$$

The following theorem presents a sufficient condition for the exponentially mean square stability of the sliding motion dynamics (2.14).

Theorem 2.1 *The system (2.14) is exponentially stable in the mean square if there exist a positive definite matrix $P > 0$, positive scalars $\epsilon > 0$ and $\varphi_1 > 0$, such that*

$$\begin{bmatrix} -P - \varphi_1\tilde{U} & -\varphi_1\bar{U} & 2A^\mathrm{T}P & 2A^\mathrm{T}PB & E_2^\mathrm{T}P & 0 & \epsilon N^\mathrm{T} \\ -\varphi_1\bar{U}^\mathrm{T} & 2E_1^\mathrm{T}PE_1 - \varphi_1 I & 0 & 0 & 0 & 0 & 0 \\ 2PA & 0 & -P & 0 & 0 & 2PM & 0 \\ 2B^\mathrm{T}PA & 0 & 0 & -B^\mathrm{T}PB & 0 & 2B^\mathrm{T}PM & 0 \\ PE_2 & 0 & 0 & 0 & -P & 0 & 0 \\ 0 & 0 & 2M^\mathrm{T}P & 2M^\mathrm{T}PB & 0 & -\epsilon I & 0 \\ \epsilon N & 0 & 0 & 0 & 0 & 0 & -\epsilon I \end{bmatrix} < 0 \tag{2.22}$$

$$B^\mathrm{T}PE = 0.$$

Proof *For system (2.14), we choose the Lyapunov functional by* $\mathcal{W}(k) = x^{\mathrm{T}}(k)Px(k)$. *Then, along the trajectory, we have*

$$
\begin{aligned}
\mathbb{E}\{\Delta\mathcal{W}|x(k)\} =&\,\mathbb{E}\{\mathcal{W}(k+1)|x(k)\} - \mathcal{W}(k) \\
=&\,\mathbb{E}\{x^{\mathrm{T}}(k+1)Px(k+1)|x(k)\} - x^{\mathrm{T}}(k)Px(k) \\
=&\,\mathbb{E}\{(A_K x(k) + E_1 g(x(k)) + E_2 x(k)\omega(k))^{\mathrm{T}} \\
&\times P(A_K x(k) + E_1 g(x(k)) + E_2 x(k)\omega(k))|x(k)\} \\
&- x^{\mathrm{T}}(k)Px(k) \\
=&\,x^{\mathrm{T}}(k)(A_K^{\mathrm{T}}PA_K + E_2^{\mathrm{T}}PE_2 - P)x(k) \\
&+ g^{\mathrm{T}}(x(k))E_1^{\mathrm{T}}PE_1 g(x(k)) + 2x^{\mathrm{T}}(k)A_K^{\mathrm{T}}PE_1 g(x(k)).
\end{aligned}
\tag{2.23}
$$

By Lemma 2.2, it is easy to obtain

$$
\begin{aligned}
2x^{\mathrm{T}}(k)A_K^{\mathrm{T}}PE_1 g(x(k)) \leq &\,x^{\mathrm{T}}(k)A_K^{\mathrm{T}}PA_K x(k) \\
&+ g^{\mathrm{T}}(x(k))E_1^{\mathrm{T}}PE_1 g(x(k))
\end{aligned}
\tag{2.24}
$$

and

$$
\begin{aligned}
A_K^{\mathrm{T}}PA_K =&\,(A + \Delta A - B(B^{\mathrm{T}}PB)^{-1}B^{\mathrm{T}}P(A + \Delta A))^{\mathrm{T}} \\
&\times P(A + \Delta A - B(B^{\mathrm{T}}PB)^{-1}B^{\mathrm{T}}P(A + \Delta A)) \\
\leq&\,2(A + \Delta A)^{\mathrm{T}}P(A + \Delta A) + 2\bar{A}^{\mathrm{T}}P\bar{A}
\end{aligned}
\tag{2.25}
$$

where $\bar{A}$ *is defined as* $\bar{A} \triangleq -B(B^{\mathrm{T}}PB)^{-1}B^{\mathrm{T}}P(A+\Delta A)$. *To this end, we have*

$$
\begin{aligned}
A_K^{\mathrm{T}}PA_K \leq &\,2(A + \Delta A)^{\mathrm{T}}P(A + \Delta A) \\
&+ 2(A + \Delta A)^{\mathrm{T}}PB(B^{\mathrm{T}}PB)^{-1}B^{\mathrm{T}}P(A + \Delta A).
\end{aligned}
\tag{2.26}
$$

Notice that, when $z = 0$, *inequality (2.6) is equivalent to*

$$
\begin{aligned}
\left[\begin{array}{c} x(k) \\ g(x(k)) \end{array} \right]^{\mathrm{T}} \left[\begin{array}{cc} \tilde{U} & \bar{U} \\ \bar{U}^{\mathrm{T}} & I \end{array} \right] \left[\begin{array}{c} x(k) \\ g(x(k)) \end{array} \right] \leq 0, \\
\left[\begin{array}{c} x(k) \\ h(x(k)) \end{array} \right]^{\mathrm{T}} \left[\begin{array}{cc} \tilde{V} & \bar{V} \\ \bar{V}^{\mathrm{T}} & I \end{array} \right] \left[\begin{array}{c} x(k) \\ h(x(k)) \end{array} \right] \leq 0.
\end{aligned}
\tag{2.27}
$$

Therefore, for some $\varphi_1 > 0$,

$$
\begin{aligned}
&\mathbb{E}\{\Delta\mathcal{W}|x(k)\} \\
\leq&\,\mathbb{E}\{\Delta\mathcal{W}|x(k)\} - \varphi_1 \left[\begin{array}{c} x(k) \\ g(x(k)) \end{array} \right]^{\mathrm{T}} \left[\begin{array}{cc} \tilde{U} & \bar{U} \\ \bar{U}^{\mathrm{T}} & I \end{array} \right] \left[\begin{array}{c} x(k) \\ g(x(k)) \end{array} \right] \\
=&\,\xi^{\mathrm{T}}(k) \left[\begin{array}{cc} 2A_K^{\mathrm{T}}PA_K + E_2^{\mathrm{T}}PE_2 - P - \varphi_1\tilde{U} & -\varphi_1\bar{U} \\ -\varphi_1\bar{U}^{\mathrm{T}} & 2E_1^{\mathrm{T}}PE_1 - \varphi_1 I \end{array} \right] \xi(k)
\end{aligned}
\tag{2.28}
$$

where

$$
\xi(k) \triangleq \left[\begin{array}{c} x(k) \\ g(x(k)) \end{array} \right].
$$

By Schur Complement,

$$\begin{bmatrix} 2A_K^T P A_K + E_2^T P E_2 - P - \varphi_1 \widetilde{U} & -\varphi_1 \bar{U} \\ -\varphi_1 \bar{U}^T & 2E_1^T P E_1 - \varphi_1 I \end{bmatrix} < 0$$

$$\Longleftrightarrow \begin{bmatrix} -P - \varphi_1 \widetilde{U} & -\varphi_1 \bar{U} & \sqrt{2} A_K^T P & E_2^T P \\ -\varphi_1 \bar{U}^T & 2E_1^T P E_1 - \varphi_1 I & 0 & 0 \\ \sqrt{2} P A_K & 0 & -P & 0 \\ P E_2 & 0 & 0 & -P \end{bmatrix} < 0. \qquad (2.29)$$

It is easy to see that (2.29) is implied by

$$\begin{bmatrix} -P - \varphi_1 \widetilde{U} & -\varphi_1 \bar{U} & 2(A + \Delta A)^T P \\ -\varphi_1 \bar{U}^T & 2E_1^T P E_1 - \varphi_1 I & 0 \\ 2P(A + \Delta A) & 0 & -P \\ 2B^T P(A + \Delta A) & 0 & 0 \\ P E_2 & 0 & 0 \end{bmatrix}$$

$$\left. \begin{matrix} 2(A + \Delta A)^T P B & E_2^T P \\ 0 & 0 \\ 0 & 0 \\ -B^T P B & 0 \\ 0 & -P \end{matrix} \right] < 0. \qquad (2.30)$$

Now, rewrite matrix inequality (2.30) into the following form:

$$\begin{bmatrix} -P - \varphi_1 \widetilde{U} & -\varphi_1 \bar{U} & 2A^T P & 2A^T P B & E_2^T P \\ -\varphi_1 \bar{U}^T & 2E_1^T P E_1 - \varphi_1 I & 0 & 0 & 0 \\ 2P A & 0 & -P & 0 & 0 \\ 2B^T P A & 0 & 0 & -B^T P B & 0 \\ P E_2 & 0 & 0 & 0 & -P \end{bmatrix} \qquad (2.31)$$

$$+ \bar{M} F \bar{N} + \bar{N}^T F^T \bar{M}^T < 0$$

where

$$\bar{M} = \begin{bmatrix} 0 & 0 & 2M^T P & 2M^T P B & 0 \end{bmatrix}^T,$$
$$\bar{N} = \begin{bmatrix} N & 0 & 0 & 0 & 0 \end{bmatrix}.$$

By Lemma 2.4, we can see that (2.31) is true if and only if there exists a $\epsilon > 0$ such that:

$$\begin{bmatrix} -P - \varphi_1 \widetilde{U} & -\varphi_1 \bar{U} & 2A^T P & 2A^T P B & E_2^T P \\ -\varphi_1 \bar{U}^T & 2E_1^T P E_1 - \varphi_1 I & 0 & 0 & 0 \\ 2P A & 0 & -P & 0 & 0 \\ 2B^T P A & 0 & 0 & -B^T P B & 0 \\ P E_2 & 0 & 0 & 0 & -P \end{bmatrix} \qquad (2.32)$$

$$+ \epsilon^{-1} \bar{M} \bar{M}^T + \epsilon N^T \bar{N} < 0.$$

It follows again from Lemma 2.4 that (2.32) is equivalent to (2.22). Then, we have from (2.28) that $\mathbb{E}\{\Delta\mathcal{W}|x(k)\} < 0$ *which indicates the sliding motion dynamics (2.14) is asymptotically mean square stable. Moreover, from (2.22), it is seen that*

$$\Omega \triangleq \begin{bmatrix} 2A_K^{\mathrm{T}} P A_K + E_2^{\mathrm{T}} P E_2 - P - \varphi_1 \tilde{U} & -\varphi_1 \bar{U} \\ -\varphi_1 \bar{U}^{\mathrm{T}} & 2E_1^{\mathrm{T}} P E_1 - \varphi_1 I \end{bmatrix} < 0, \quad (2.33)$$

from which we know that there must exist a sufficiently small scalar α *satisfying* $0 < \alpha < \lambda_{\max}(P)$ *such that* $\Omega < -\alpha I$. *Therefore, it follows that*

$$\begin{aligned} \mathbb{E}\{\Delta\mathcal{W}|x(k)\} = \mathbb{E}\{\mathcal{W}(k+1)|x(k)\} - \mathcal{W}(k) \\ \leq -\alpha\xi^{\mathrm{T}}(k)\xi(k) \\ \leq -\alpha x^{\mathrm{T}}(k)x(k) \\ \leq -\frac{\alpha}{\lambda_{\max}(P)}\mathcal{W}(k). \end{aligned} \quad (2.34)$$

Then, the exponentially mean square stability of system (2.14) can be verified immediately from Lemma 2.2 and Definition 2.1. The proof is complete.

By means of LMI, the following theorem establishes a unified framework within which the exponentially mean square stability can be guaranteed together with the pre-specified $\mathcal{H}_\infty$ noise attenuation level.

Theorem 2.2 *Consider the system (2.14). For the pre-specified* $\mathcal{H}_\infty$ *noise attenuation level* $\gamma > 0$, *if there exist a positive definite matrix* $P > 0$, *positive scalars* $\epsilon > 0$, $\varphi_1 > 0$ *and* $\varphi_2 > 0$ *satisfying*

$$\begin{bmatrix} \Upsilon_{11} & -\varphi_1\bar{U} & -\varphi_2\bar{V} + \bar{\Theta}C^{\mathrm{T}}E_3 \\ * & -\varphi_1 I + 2E_1^{\mathrm{T}}PE_1 & 0 \\ * & * & -\varphi_2 I + E_3^{\mathrm{T}}E_3 \\ * & * & * \\ * & * & * \\ * & * & * \\ * & * & * \\ * & * & * \end{bmatrix}$$

$$\begin{matrix} \bar{\Theta}C^{\mathrm{T}}D & 2A^{\mathrm{T}}P & 2A^{\mathrm{T}}PB & 0 & \epsilon N^{\mathrm{T}} \\ 0 & 0 & 0 & 0 & 0 \\ E_3^{\mathrm{T}}D & 0 & 0 & 0 & 0 \\ D^{\mathrm{T}}D - \gamma^2 I & 0 & 0 & 0 & 0 \\ * & -P & 0 & 2PM & 0 \\ * & * & -B^{\mathrm{T}}PB & 2B^{\mathrm{T}}PM & 0 \\ * & * & * & -\epsilon I & 0 \\ * & * & * & * & -\epsilon I \end{matrix} \quad < \quad 0 \quad (2.35)$$

$$B^{\mathrm{T}}PE = 0 \quad (2.36)$$

where

$$\Upsilon_{11} = E_2^{\mathrm{T}}PE_2 + \bar{\Theta}C^{\mathrm{T}}C\bar{\Theta} + \sum_{i=1}^{n} \sigma_i^2 C_i^{\mathrm{T}}C_i - P - \varphi_1\tilde{U} - \varphi_2\tilde{V},$$

then the system (2.14) is exponentially mean square stable and, meanwhile, the $\mathcal{H}_\infty$ performance is achieved.

Proof *It is obvious that Theorem 2.2 implies Theorem 2.1, and therefore the system (2.14) is exponentially mean square stable.*

Next, for any $\nu(k) \neq 0$, we have

$$
\begin{aligned}
&\mathbb{E}\{y^\mathrm{T}(k)y(k)\} \\
=&\mathbb{E}\{(C\Theta x(k) + E_3 h(x(k)) + D\nu(k))^\mathrm{T}(C\Theta x(k) \\
&+ E_3 h(x(k)) + D\nu(k))\} \\
=&\mathbb{E}\{x^\mathrm{T}(k)\Theta C^\mathrm{T} C\Theta x(k) + 2x^\mathrm{T}(k)\Theta C^\mathrm{T} E_3 h(x(k)) \\
&+ 2x^\mathrm{T}(k)\Theta C^\mathrm{T} D\nu(k)\} + h^\mathrm{T}(x(k))E_3^\mathrm{T} E_3 h(x(k)) \\
&+ \nu^\mathrm{T}(k)D^\mathrm{T} D\nu(k) + 2\nu^\mathrm{T}(k)D^\mathrm{T} E_3 h(x(k)).
\end{aligned}
\tag{2.37}
$$

Defining $\widetilde{\Theta} \triangleq \Theta - \bar{\Theta}$, we obtain

$$
\begin{aligned}
&\mathbb{E}\{x^\mathrm{T}(k)\Theta C^\mathrm{T} C\Theta x(k)\} \\
=&\mathbb{E}\{x^\mathrm{T}(k)(\bar{\Theta} + \widetilde{\Theta})C^\mathrm{T} C(\bar{\Theta} + \widetilde{\Theta})x(k)\} \\
=&x^\mathrm{T}(k)\bar{\Theta} C^\mathrm{T} C\bar{\Theta} x(k) + \mathbb{E}\{2x^\mathrm{T}(k)\bar{\Theta} C^\mathrm{T} C\widetilde{\Theta} x(k)\} \\
&+ \mathbb{E}\{x^\mathrm{T}(k)\widetilde{\Theta} C^\mathrm{T} C\widetilde{\Theta} x(k)\} \\
=&x^\mathrm{T}(k)\bar{\Theta} C^\mathrm{T} C\bar{\Theta} x(k) + x^\mathrm{T}(k)(\sum_{i=1}^{n} \sigma_i^2 C_i^\mathrm{T} C_i)x(k)
\end{aligned}
\tag{2.38}
$$

and

$$
\begin{aligned}
&\mathbb{E}\{\mathcal{W}(k+1)|x(k)\} - \mathcal{W}(k) + \mathbb{E}\{y^\mathrm{T}(k)y(k)\} - \gamma^2\mathbb{E}\{\nu^\mathrm{T}(k)\nu(k)\} \\
=&\mathbb{E}\{x^\mathrm{T}(k)(A_K^\mathrm{T} PA_K + E_2^\mathrm{T} PE_2 + \bar{\Theta} C^\mathrm{T} C\bar{\Theta} + \sum_{i=1}^{n} \sigma_i^2 C_i^\mathrm{T} C_i - P)x(k) \\
&+ 2x^\mathrm{T}(k)A_K^\mathrm{T} PE_1 g(x(k)) + g^\mathrm{T}(x(k))E_1^\mathrm{T} PE_1 g(x(k)) \\
&+ 2x^\mathrm{T}(k)\bar{\Theta} C^\mathrm{T} E_3 h(x(k)) + 2x^\mathrm{T}(k)\bar{\Theta} C^\mathrm{T} D\nu(k) \\
&+ h^\mathrm{T}(x(k))E_3^\mathrm{T} E_3 h(x(k)) + \nu^\mathrm{T}(k)D^\mathrm{T} D\nu(k) \\
&+ 2\nu^\mathrm{T}(k)D^\mathrm{T} E_3 h(x(k))\} - \gamma^2\mathbb{E}\{\nu^\mathrm{T}(k)\nu(k)\}
\end{aligned}
\tag{2.39}
$$

Taking (2.24) and (2.27) into consideration, for some $\varphi_1 > 0$ and $\varphi_2 > 0$, *we have*

$$\mathbb{E}\{\mathcal{W}(k+1)|x(k)\} - \mathcal{W}(k) + \mathbb{E}\{y^{\mathrm{T}}(k)y(k)\} - \gamma^2\mathbb{E}\{\nu^{\mathrm{T}}(k)\nu(k)\}$$

$$\leq \mathbb{E}\{\mathcal{W}(k+1)|x(k)\} - \mathcal{W}(k) + \mathbb{E}\{y^{\mathrm{T}}(k)y(k)\} - \gamma^2\mathbb{E}\{\nu^{\mathrm{T}}(k)\nu(k)\}$$

$$- \varphi_1 \begin{bmatrix} x(k) \\ g(x(k)) \end{bmatrix}^{\mathrm{T}} \begin{bmatrix} \tilde{U} & \bar{U} \\ \bar{U}^{\mathrm{T}} & I \end{bmatrix} \begin{bmatrix} x(k) \\ g(x(k)) \end{bmatrix} \tag{2.40}$$

$$- \varphi_2 \begin{bmatrix} x(k) \\ h(x(k)) \end{bmatrix}^{\mathrm{T}} \begin{bmatrix} \tilde{V} & \bar{V} \\ \bar{V}^{\mathrm{T}} & I \end{bmatrix} \begin{bmatrix} x(k) \\ h(x(k)) \end{bmatrix}$$

$$\triangleq \mathbb{E}\{\eta(k)^{\mathrm{T}}\Upsilon\eta(k)\}.$$

Here,

$$\Upsilon \triangleq \begin{bmatrix} \Upsilon_{11} + 2A_K^{\mathrm{T}}PA_K & -\varphi_1\bar{U} \\ * & -\varphi_1 I + 2E_1^{\mathrm{T}}PE_1 \\ * & * \\ * & * \end{bmatrix}$$

$$\begin{matrix} -\varphi_2\bar{V} + \bar{\Theta}C^{\mathrm{T}}E_3 & \bar{\Theta}C^{\mathrm{T}}D \\ 0 & 0 \\ -\varphi_2 I + E_3^{\mathrm{T}}E_3 & E_3^{\mathrm{T}}D \\ * & D^{\mathrm{T}}D - \gamma^2 I \end{matrix} \Bigg],$$

$$\eta(k) \triangleq \begin{bmatrix} x^{\mathrm{T}}(k) & g^{\mathrm{T}}(x(k)) & h^{\mathrm{T}}(x(k)) & \nu^{\mathrm{T}}(k) \end{bmatrix}^{\mathrm{T}}.$$

Using Schur Complement, $\Upsilon < 0$ *is true if*

$$\begin{bmatrix} \Upsilon_{11} & -\varphi_1\bar{U} & -\varphi_2\bar{V} + \bar{\Theta}C^{\mathrm{T}}E_3 \\ * & -\varphi_1 I + E_1^{\mathrm{T}}PE_1 & 0 \\ * & * & -\varphi_2 I + E_3^{\mathrm{T}}E_3 \\ * & * & * \\ * & * & * \\ * & * & * \end{bmatrix}$$

$$\begin{matrix} \bar{\Theta}C^{\mathrm{T}}D & 2(A+\Delta A)^{\mathrm{T}}P & 2(A+\Delta A)^{\mathrm{T}}PB \\ 0 & 0 & 0 \\ E_3^{\mathrm{T}}D & 0 & 0 \\ D^{\mathrm{T}}D - \gamma^2 I & 0 & 0 \\ * & -P & 0 \\ * & * & -B^{\mathrm{T}}PB \end{matrix} \Bigg] < 0 \tag{2.41}$$

which, by Lemma 2.4, is equivalent to (2.35), and therefore

$$\mathbb{E}\{\mathcal{W}(k+1)|x(k)\} - \mathcal{W}(k) + \mathbb{E}\{y^{\mathrm{T}}(k)y(k)\} - \gamma^2\mathbb{E}\{\nu^{\mathrm{T}}(k)\nu(k)\} < 0. \tag{2.42}$$

Next, taking the sum on both sides of (2.42) from 0 to ∞ with respect to k leads to

$$\sum_{k=0}^{\infty} \Big[\mathbb{E}\{\mathcal{W}(k+1)|x(k)\} - \mathcal{W}(k)$$

$$+ \mathbb{E}\{y^{\mathrm{T}}(k)y(k)\} - \gamma^2\mathbb{E}\{\nu^{\mathrm{T}}(k)\nu(k)\} \Big] < 0, \tag{2.43}$$

or

$$\sum_{k=0}^{\infty} \mathbb{E}\{\|y(k)\|^2\} < \gamma^2 \sum_{k=0}^{\infty} \mathbb{E}\{\|\nu(k)\|^2\} + \mathcal{W}(0) - \mathcal{W}(\infty). \tag{2.44}$$

Since $x(0) = 0$ *and the system* (2.14) *is exponentially mean square stable, we can easily obtain*

$$\sum_{k=0}^{\infty} \mathbb{E}\{\|y(k)\|^2\} < \gamma^2 \sum_{k=0}^{\infty} \mathbb{E}\{\|\nu(k)\|^2\}, \tag{2.45}$$

which completes the proof.

Computational Algorithm

Notice that the condition in Theorem 2.2 is presented as the feasibility problem of an LMI with an equality constraint. By means of the proposed method in [167], as the condition $B^{\mathrm{T}} PE = 0$ is equivalent to $\mathrm{tr}[(B^{\mathrm{T}} PE)^{\mathrm{T}} B^{\mathrm{T}} PE] = 0$, we first introduce the condition $(B^{\mathrm{T}} PE)^{\mathrm{T}} B^{\mathrm{T}} PE \le \phi I$. By Schur Complement, the condition can be expressed as

$$\begin{bmatrix} -\phi I & E^{\mathrm{T}} PB \\ B^{\mathrm{T}} PE & -I \end{bmatrix} < 0. \tag{2.46}$$

Hence, the original nonconvex feasibility problem can be converted into the following minimization problem:

$$\min \phi \quad \text{subject to} \quad (2.35) \text{ and } (2.46). \tag{2.47}$$

If this minimum equals zero, the solutions will satisfy the LMI (2.35) with the equality $B^{\mathrm{T}} PE = 0$. Thus, the exponentially mean square stability and $\mathcal{H}_\infty$ performance of system (2.14) are simultaneously achieved.

Reaching Condition Analysis

In this subsection, we will synthesize a sliding mode controller, with the pre-specified stochastic switching function (2.10) and sliding surface (2.12), to meet the discrete-time sliding mode reaching condition. That is to say, the trajectory of (2.1) starting from any initial state is globally driven onto the sliding surface (2.12) in finite time with probability 1, and then results in a sliding motion within a band called quasi-sliding mode band (QSMB) [212], along the sliding surface in the subsequent time.

To begin with, since the system parameter uncertainty ΔA and external disturbance $f(x(k))$ are both assumed to be bounded, $D_a \triangleq B^{\mathrm{T}} P \Delta A x(k)$ and $D_f \triangleq B^{\mathrm{T}} PB f(x(k))$ will also be bounded. Denote d_a^i and d_f^i as the ith element in D_a and D_f, respectively. Suppose the lower and upper bounds on

D_a and D_f are known and given as follows:

$$d_{aL}^i \leq d_a^i \leq d_{aU}^i,$$
$$d_{fL}^i \leq d_f^i \leq d_{fU}^i, \qquad i = 1, 2, \cdots, m \tag{2.48}$$

where d_{aL}^i, d_{aU}^i, d_{fL}^i and d_{fU}^i are all known constants. Furthermore, we denote

$$\bar{D}_a = \begin{bmatrix} \bar{d}_a^1 & \bar{d}_a^2 & \cdots & \bar{d}_a^m \end{bmatrix}^{\mathrm{T}},$$
$$\bar{d}_a^i = \frac{d_{aU}^i + d_{aL}^i}{2},$$
$$\tilde{D}_a = \mathrm{diag}\left\{ \tilde{d}_a^1, \tilde{d}_a^2, \cdots, \tilde{d}_a^m \right\},$$
$$\tilde{d}_a^i = \frac{d_{aU}^i - d_{aL}^i}{2},$$
$$\bar{D}_f = \begin{bmatrix} \bar{d}_f^1 & \bar{d}_f^2 & \cdots & \bar{d}_f^m \end{bmatrix}^{\mathrm{T}}, \tag{2.49}$$
$$\bar{d}_f^i = \frac{d_{fU}^i + d_{fL}^i}{2},$$
$$\tilde{D}_f = \mathrm{diag}\left\{ \tilde{d}_f^1, \tilde{d}_f^2, \cdots, \tilde{d}_f^m \right\},$$
$$\tilde{d}_f^i = \frac{d_{fU}^i - d_{fL}^i}{2}, \qquad i \qquad = 1, 2, \cdots, m.$$

Remark 2.5 *We should point out that the assumption on the upper and lower bounds of D_a and D_f are standard for discrete-time SMC; see [212] and the references therein. Besides, the bounds of both D_a and D_f might be time-varying or dependent on state $x(k)$, which we will show in Section 2.3.*

Next, we aim to improve the reaching condition proposed in [212] by proposing the following form for system (2.1) with the sliding surface (2.12):

$$\mathbb{E}\{\Delta s_i(k)\} = \mathbb{E}\{s_i(k+1) - s_i(k)\}$$
$$\leq -\rho\lambda_i \cdot \mathrm{sgn}[\mathbb{E}\{s_i(k)\}] - \rho q_i \mathbb{E}\{s_i(k)\} \quad \text{if } \mathbb{E}\{s_i(k)\} > 0$$
$$\mathbb{E}\{\Delta s_i(k)\} = \mathbb{E}\{s_i(k+1) - s_i(k)\} \tag{2.50}$$
$$\geq -\rho\lambda_i \cdot \mathrm{sgn}[\mathbb{E}\{s_i(k)\}] - \rho q_i \mathbb{E}\{s_i(k)\} \quad \text{if } \mathbb{E}\{s_i(k)\} < 0$$

where ρ represents the sampling period, $\lambda_i > 0$ and $q_i > 0$ $(i = 1, 2, \cdots, m)$ are properly chosen scalars satisfying $0 < 1 - \rho q_i < 1$, $\forall i \in \{1, 2, \cdots, m\}$. We also can rewrite (2.50) into a compact form as follows:

$$\mathbb{E}\{\Delta s(k)\} = \mathbb{E}\{s(k+1) - s(k)\}$$
$$\leq -\rho\Lambda \cdot \mathrm{sgn}[\mathbb{E}\{s(k)\}] - \rho Q \mathbb{E}\{s(k)\} \qquad \text{if } \mathbb{E}\{s(k)\} > 0$$
$$\mathbb{E}\{\Delta s(k)\} = \mathbb{E}\{s(k+1) - s(k)\} \tag{2.51}$$
$$\geq -\rho\Lambda \cdot \mathrm{sgn}[\mathbb{E}\{s(k)\}] - \rho Q \mathbb{E}\{s(k)\} \qquad \text{if } \mathbb{E}\{s(k)\} < 0$$

where

$$\Lambda = \text{diag}\{\lambda_1, \lambda_2, \cdots, \lambda_m\} \in \mathbb{R}^{m \times m},$$
$$Q = \text{diag}\{q_1, q_2, \cdots, q_m\} \in \mathbb{R}^{m \times m}.$$

Now we are ready to give the design technique of the robust SMC controller.

Theorem 2.3 *Consider the uncertain nonlinear stochastic system* (2.1) *with the stochastic sliding surface* (2.12) *where P is the solution to* (2.35)-(2.36). *If the SMC law is given as*

$$
\begin{aligned}
u(k) = & -(G\bar{\Theta}B)^{-1}(\rho\Lambda \cdot \text{sgn}[G\bar{\Theta}x(k)] + (\rho Q - I)G\bar{\Theta}x(k) \\
& + G\bar{\Theta}Ax(k) + (\bar{D}_a + \bar{D}_f) + (\tilde{D}_a + \tilde{D}_f)\text{sgn}[G\bar{\Theta}x(k)]),
\end{aligned}
\tag{2.52}
$$

then the state trajectories of the system (2.1) *are driven onto the pre-specified sliding surface* (2.10) *with probability 1.*

Proof *By* (2.52), *with the switching function defined in* (2.10), *we can easily obtain*

$$
\begin{aligned}
\mathbb{E}\{\Delta s(k)\} = & \mathbb{E}\{s(k+1) - s(k)\} \\
= & \mathbb{E}\{G\Theta x(k+1) - G\Theta x(k))\} \\
= & G\bar{\Theta}((A + \Delta A)x(k) + Bu(k) + Bf(x(k)) - x(k)) \\
= & G\bar{\Theta}((A + \Delta A)x(k) - x(k) + Bf(x(k))) \\
& - (\rho\Lambda \cdot \text{sgn}[G\bar{\Theta}x(k)] + (\rho Q - I)G\bar{\Theta}x(k) + G\bar{\Theta}Ax(k) \\
& + (\bar{D}_a + \bar{D}_f) + (\tilde{D}_a + \tilde{D}_f)\text{sgn}[G\bar{\Theta}x(k)]) \\
= & B^{\mathrm{T}}P\Delta Ax(k) + B^{\mathrm{T}}PBf(x(k)) \\
& - \rho\Lambda \cdot \text{sgn}[B^{\mathrm{T}}Px(k)] - \rho Q B^{\mathrm{T}}Px(k) \\
& - (\bar{D}_a + \bar{D}_f) - (\tilde{D}_a + \tilde{D}_f)\text{sgn}[B^{\mathrm{T}}Px(k)] \\
= & -\rho\Lambda \cdot \text{sgn}[\mathbb{E}\{s(k)\}] - \rho Q\mathbb{E}\{s(k)\} \\
& + B^{\mathrm{T}}P\Delta Ax(k) - (\bar{D}_a + \tilde{D}_a\text{sgn}[\mathbb{E}\{s(k)\}]) \\
& + B^{\mathrm{T}}PBf(x(k)) - (\bar{D}_f + \tilde{D}_f\text{sgn}[\mathbb{E}\{s(k)\}])
\end{aligned}
\tag{2.53}
$$

and

$$
\mathbb{E}\{s(k)\} < 0 \Longrightarrow
\begin{cases}
B^{\mathrm{T}}P\Delta Ax(k) \geq \bar{D}_a + \tilde{D}_a\text{sgn}[\mathbb{E}\{s(k)\}] \\
B^{\mathrm{T}}PBf(x(k)) \geq \bar{D}_f + \tilde{D}_f\text{sgn}[\mathbb{E}\{s(k)\}]
\end{cases}
\tag{2.54}
$$
$$\Longrightarrow \mathbb{E}\{\Delta s(k)\} \geq -\rho\Lambda \cdot \text{sgn}[\mathbb{E}\{s(k)\}] - \rho Q\mathbb{E}\{s(k)\}.$$

Similarly, we can obtain

$$\mathbb{E}\{s(k)\} > 0 \Longrightarrow \mathbb{E}\{\Delta s(k)\} \leq -\rho\Lambda \cdot \text{sgn}[\mathbb{E}\{s(k)\}] - \rho Q\mathbb{E}\{s(k)\}. \tag{2.55}$$

Therefore, the reaching condition (2.51) *for discrete-time sliding mode is satisfied. In other words, the trajectory of system* (2.1) *will be, with probability* 1, *globally driven on the pre-specified sliding surface in finite time and result in a non-increasing sliding motion within the quasi-sliding mode band afterwards. The proof ends.*

Remark 2.6 *We point out that it is not difficult to extend the present results to more general systems that include polytopic parameter uncertainties, stochastic disturbances and constant or time-varying time delays by using the approach proposed and the LMI framework developed. The reason why we discuss the simplified system* (2.1)-(2.2) *is to make our theory more understandable and also to avoid unnecessarily complicated notations.*

Some Discussions

First, let us discuss the issue of worst-case analysis of the robustness. In this chapter, we have considered three kinds of "perturbations", i.e., the parameter uncertainties ΔA, the external disturbances $\nu(k)$ and the packet losses. For ΔA, it has been shown that, as long as the norm-bounded condition (2.4) holds, our main results are true no matter how ΔA varies within the bounded set. In this sense, we have actually dealt with the worst-case analysis with respect to ΔA. For $\nu(k)$, we have introduced the requirement (Q2), $\mathcal{H}_\infty$ performance constraint, to account for the disturbance rejection attenuation level. $\mathcal{H}_\infty$ performance, as is well known, can be understood as the worst-case property ("best out of the worst") as long as the disturbance $\nu(k)$ has bounded energy. Therefore, our main results are true; i.e., the disturbance rejection attenuation level is guaranteed no matter how $\nu(k)$ varies within an energy-bounded set.

Second, let us discuss the practical stability/boundedness issue. In this chapter, the proposed switching function is actually a *stochastic difference equation*. In case there is a stochastic disturbance, it would be more reasonable to deal with the stability ***in a probabilistic way*** rather than the absolute stability. In fact, one of the novelties of this chapter lies in the stochastic analysis of the sliding mode behavior. Although the stability in probability 1 (considered in this chapter) is weaker than the absolute stability, it has been widely used in the stochastic control area, see e.g. [7, 168, 239]. For example, in [239], the stability of the stochastic system is introduced in a probabilistic way, and the designed controller is also said to be able to stabilize the system in a probabilistic sense. Accordingly, in this chapter, we propose to discuss the sliding mode control problem for the stochastic nonlinear system in a probabilistic way (with probability 1).

In order to show how the stability performances are influenced by the stochastic factors, we have added two figures in the simulation part to illustrate the worst-case (that is, all the data packets are lost during the sampling;

hence no valid signal can be used for feedback) response. It can be seen that the system is not stable when the states are completely lost.

2.3 An Illustrative Example

In this section, we present an illustrative example to demonstrate the effectiveness of the proposed algorithm. The nominal system matrix A is taken from the model of an F-404 aircraft engine system in [53]. Note that this example is actually a special case of the physical models studied in [25, 53]. Moreover, in order to make the model more realistic and more close to real-world engineering practices, we add the stochastic noises, the data packet loss and the external disturbances caused by the complex and time-varying working conditions in the system. After discretization, the system is as follows:

$$
\left\{
\begin{aligned}
x(k+1) &= \left(\begin{bmatrix} 0.0307 & 0 & 0.0557 \\ 0.0333 & 0.2466 & -0.0091 \\ 0.0071 & 0 & 0.0130 \end{bmatrix} \right. \\
&\quad + \begin{bmatrix} 0.01 \\ 0.02 \\ 0 \end{bmatrix} \sin(0.6k) \begin{bmatrix} 0 & 0.01 & 0 \end{bmatrix} \left. \right) x(k) \\
&\quad + \begin{bmatrix} 0.1817 & 0.4286 \\ 0.1597 & 0.0793 \\ 0.1138 & 0.0581 \end{bmatrix} \big(u(k) + f(x(k)) \big) \\
&\quad + \begin{bmatrix} 0.03 & 0 & -0.01 \\ 0.02 & 0.03 & 0 \\ 0.04 & 0.05 & -0.01 \end{bmatrix} g(x(k)) \\
&\quad + \begin{bmatrix} 0.015 & 0 & -0.01 \\ 0.01 & 0.015 & 0 \\ 0.02 & 0.025 & -0.01 \end{bmatrix} x(k)\omega(k), \\
y(k) &= \begin{bmatrix} 0.2 & 0 & -0.1 \\ 0.1 & 0.15 & 0 \end{bmatrix} \Theta x(k) \\
&\quad + \begin{bmatrix} -0.01 & 0 & 0.03 \\ 0.01 & 0.02 & 0 \end{bmatrix} h(x(k)) + \begin{bmatrix} 0.015 \\ 0.02 \end{bmatrix} \nu(k).
\end{aligned}
\right.
\tag{2.56}
$$

Let

$$
\begin{aligned}
f(x(k)) &= \begin{bmatrix} 0.5\sin(x_1(k)) \\ 0.6\cos(x_3(k)) \end{bmatrix}, \\
g(x(k)) &= 0.5(U_1 + U_2)x(k) + 0.5(U_2 - U_1)\sin(x(k))x(k), \\
h(x(k)) &= 0.5(V_1 + V_2)x(k) + 0.5(V_2 - V_1)\cos(x(k))x(k),
\end{aligned}
\tag{2.57}
$$

where

$$\sin(x(k)) \triangleq \text{diag}\{\sin(x_1(k)), \sin(x_2(k)), \sin(x_3(k))\},$$
$$\cos(x(k)) \triangleq \text{diag}\{\cos(x_1(k)), \cos(x_2(k)), \cos(x_3(k))\},$$
$$U_1 = \text{diag}\{0.1, 0.2, 0.5\},$$
$$U_2 = \text{diag}\{0.1, 0.6, 0.7\},\qquad\text{(2.58)}$$
$$V_1 = \text{diag}\{0.3, 0.2, 0.8\},$$
$$V_2 = \text{diag}\{0.4, 0.5, 0.6\}.$$

In addition, we assume the probabilistic density functions of θ_1, θ_2 and θ_3 in $[0\ \ 1]$ are described by

$$\varrho_1(s_1) = \begin{cases} 0.8 & s_1 = 0 \\ 0.1 & s_1 = 0.5 \\ 0.1 & s_1 = 1, \end{cases}$$

$$\varrho_2(s_2) = \begin{cases} 0.7 & s_2 = 0 \\ 0.2 & s_2 = 0.5 \\ 0.1 & s_2 = 1, \end{cases} \qquad\text{(2.59)}$$

$$\varrho_3(s_3) = \begin{cases} 0 & s_3 = 0 \\ 0.1 & s_3 = 0.5 \\ 0.9 & s_3 = 1, \end{cases}$$

from which the expectations and variances can be easily calculated as $\mu_1 = 0.15$, $\sigma_1^2 = 0.1025$, $\mu_2 = 0.2$, $\sigma_2^2 = 0.11$, $\mu_3 = 0.95$ and $\sigma_3^2 = 0.0225$.

Choosing $\gamma = 0.8$ and using Matlab LMI Toolbox to solve problem (2.47), we have

$$P = \begin{bmatrix} 0.075173 & 0.020922 & -0.060289 \\ 0.020922 & 0.18556 & -0.091766 \\ -0.060289 & -0.091766 & 0.19063 \end{bmatrix}$$

and $\phi = 9.000011 \times 10^{-7}$ (hence the constraint $B^{\mathrm{T}} PE = 0$ is satisfied). Choose $\rho = 0.05$, $\lambda_j = 1$ and $q_j = 1$ $(j = 1, 2)$. Moreover, In order to design the explicit SMC controller, we suppose $B^{\mathrm{T}} P \Delta Ax(k)$ and $B^{\mathrm{T}} PBf(x(k))$ are bounded by the following conditions:

$$d_{aL}^i = -\|B^{\mathrm{T}} PM\|\|Nx(k)\|,$$
$$d_{aU}^i = \|B^{\mathrm{T}} PM\|\|Nx(k)\|,$$
$$d_{fL}^i = -0.5\|B^{\mathrm{T}} PB\sin(x(k))\|,\qquad\text{(2.60)}$$
$$d_{fU}^i = 0.5\|B^{\mathrm{T}} PB\sin(x(k))\|.$$

Then, it follows from Theorem 2.3 that the desired SMC law can be set up with all known parameters. The simulation results are shown in Fig. 2.1 to Fig. 2.6, which confirm that the desired requirements are well achieved.

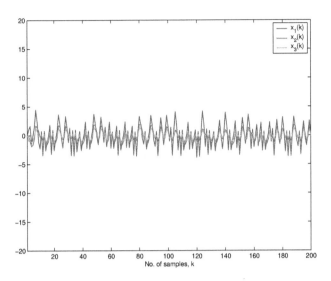

FIGURE 2.1: The state trajectories $x(k)$.

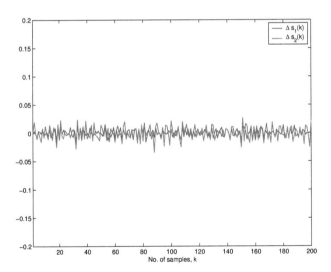

FIGURE 2.2: The signal $\Delta s(k)$.

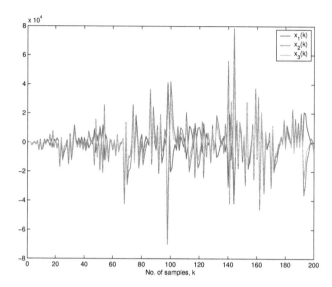

FIGURE 2.3: The state trajectories when the states are completely lost.

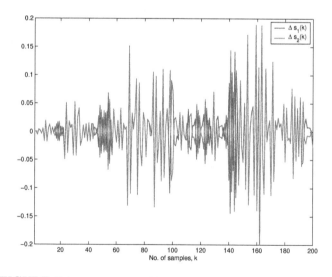

FIGURE 2.4: The signal $\Delta s(k)$ when the states are completely lost.

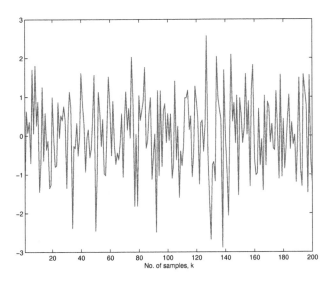

FIGURE 2.5: The noise $\omega(k)$.

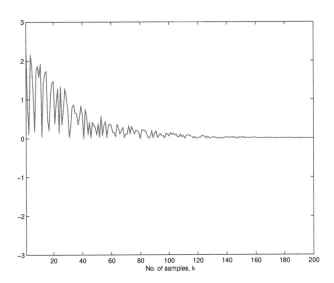

FIGURE 2.6: The disturbance $\nu(k)$.

2.4   Summary

A robust SMC design problem for a class of uncertain nonlinear stochastic systems with multiple data packet losses has been studied. Both matched and unmatched nonlinearities have been taken into consideration. The multiple data packet losses are assumed to happen in a random way, and the loss probability of each individual state variable is governed by a corresponding individual stochastic variable obeying a certain probabilistic distribution in the interval $[0 \quad 1]$. We also have introduced, for the first time, a stochastic switching function for the SMC problem of discrete-time stochastic systems. By means of LMI, a sufficient condition for the exponentially mean square stability as well as pre-specified $\mathcal{H}_\infty$ performance index of the system dynamics on the specified sliding surface has been derived. By the reaching condition proposed in this chapter, an SMC controller has been designed to globally drive the state trajectory onto the specified surface with probability 1, which gives rise to a non-increasing zigzag motion along the surface. An illustrative numerical example has been given to show the applicability and effectiveness of the proposed method in this chapter.

3

Sliding Mode Control for A Class of Nonlinear Discrete-Time Networked Systems with Multiple Stochastic Communication Delays

In the past few decades, the sliding mode control (also known as variable structure control) problem has been extensively studied and widely applied, because of its advantage of strong robustness against model uncertainties, parameter variations and external disturbances. In recent years, since most control strategies are implemented in a discrete-time setting (e.g., networked control systems), the sliding mode control (SMC) problem for discrete-time systems has gained considerable research interest.

On another research front, the past decade has witnessed the rapid development in network technologies, which has led to more and more control systems whose feedback control loops are based on networks. The usage of networks in control systems has many advantages such as low cost, reduced weight and simple installation. However, due to complex working conditions and limited bandwidth, there are certain limitations mainly including network-induced time-delays (also called communication delays) and data package loss, which will inevitably degrade system performance or even cause instability. As a result, a great deal of work has been done in order to eliminate or compensate for the effect caused by communication delays. Many researchers have studied stability and controller design problems for networked control systems (NCSs) in the presence of deterministic communication delays. Recently, due to the random and time-varying fashion of the communication delays, they have been modeled in various *probabilistic* ways.

In a networked environment, it seems sensible to apply sliding mode control strategies by taking the network-induced phenomenon into account, in order to make full use of the robustness of SMC against model uncertainties, parameter variations and external disturbances. Moreover, the model under consideration should be more comprehensive to reflect realities such as the discrete-time nature, the state-dependent stochastic disturbances (also called multiplicative noises or Itô-type noises) and the multiple randomly occurring communication delays. The possible combination of SMC and NCS would clearly have both theoretical significance and application potential. Unfortunately, the *discrete-time* sliding mode control problem for networked systems

47

has not been fully addressed, not to mention the consideration of stochastic delays and stochastic disturbances. This is mainly due to the mathematical challenges in analyzing the reaching condition for the addressed discrete-time stochastic systems. It is, therefore, the motivation of this chapter to shorten such a gap by investigating SMC schemes for a broad class of networked stochastic systems with communication delays.

This chapter aims to design a state feedback controller, by using the discrete-time sliding motion concept, such that *1)* the system state trajectory is globally driven onto the specified sliding surface which results in a non-increasing zigzag motion on the sliding surface; and *2)* the robustly exponential stability of the system is guaranteed on the specified sliding surface.

The rest of this chapter is arranged as follows. Section 3.1 formulates an uncertain nonlinear networked system with stochastic time-varying delays to be studied. In Section 3.2, a novel switching function is first designed and then an LMI-based sufficient condition is given to obtain the parameters in the proposed switching function to satisfy the robustly mean square exponential stability of the system dynamics on the sliding surface. An SMC law is synthesized in Section 3.3 to satisfy the discrete-time sliding mode reaching condition. In Section 3.4, an illustrative numerical example is provided to show the effectiveness and usefulness of the proposed approach. Section 3.5 gives our conclusions.

3.1   Problem Formulation

Consider, on a probability space $(\Omega, \mathscr{F}, \mathscr{P})$, the following uncertain nonlinear Itô-type networked system:

$$
\begin{cases}
x(k+1) = (A + \Delta A)x(k) + B(u(k) + f(x(k))) + A_d \tilde{x}(k) \\
\qquad\quad + Dg(x(k), \tilde{x}(k), k)\omega(k) \\
\tilde{x}(k) = \displaystyle\sum_{i=1}^{q} \alpha_i(k)x(k - \tau_i(k)) \\
x(k) = \varphi(k), \quad k = -d_M, -d_M + 1, \cdots, 0
\end{cases}
\tag{3.1}
$$

where $x(k) \in \mathbb{R}^n$ is the state vector, $u(k) \in \mathbb{R}^m$ is the control input, A, A_d, B, and D are known constant real-valued matrices with appropriate dimensions. $\omega(k)$ is a scalar Wiener process (Brownian Motion) on $(\Omega, \mathscr{F}, \mathscr{P})$ with

$$
\begin{aligned}
\mathbb{E}\{\omega(k)\} &= 0, \\
\mathbb{E}\{\omega^2(k)\} &= 1, \\
\mathbb{E}\{\omega(k)\omega(j)\} &= 0 \quad (k \neq j)
\end{aligned}
\tag{3.2}
$$

and $g : \mathbb{R}^n \times \mathbb{R}^n \times \mathbb{N} \to \mathbb{R}^n$ is the continuous function, and is assumed to satisfy

$$g^{\mathrm{T}}(x, y, k)g(x, y, k) \le \rho_1 x^{\mathrm{T}}x + \rho_2 y^{\mathrm{T}}y, \qquad x, y \in \mathbb{R}^n, \qquad (3.3)$$

where $\rho_1 > 0$ and $\rho_2 > 0$ are known constant scalars.

The real-valued matrix ΔA represents norm-bounded parameter uncertainty satisfying

$$\Delta A = HFE \qquad (3.4)$$

where H and E are known real constant matrices which characterize how the deterministic uncertain parameter in F enters the nominal matrix A. The matrix F, which could be time-varying, is an unknown matrix function meeting

$$F^{\mathrm{T}}F \le I. \qquad (3.5)$$

The parameter uncertainty ΔA is said to be admissible if both (3.4) and (3.5) are satisfied.

The stochastic variables $\alpha_i(k) \in \mathbb{R}$ $(i = 1, 2, \cdots, q)$ in (3.1) are mutually uncorrelated Bernoulli distributed white sequences taking values on 0 and 1 with

$$\mathcal{P}\left\{\alpha_i(k) = 1\right\} = \mathbb{E}\left\{\alpha_i(k)\right\} \triangleq \bar{\alpha}_i \qquad (3.6)$$

where $\bar{\alpha}_i$ $(i = 1, 2, \cdots, q)$ are known positive scalars. It is easy to see that

$$\begin{aligned} \mathcal{P}\left\{\alpha_i(k) = 0\right\} &= 1 - \bar{\alpha}_i, \\ \sigma_{\alpha_i}^2 \triangleq \mathbb{E}\left\{(\alpha_i(k) - \bar{\alpha}_i)^2\right\} &= (1 - \bar{\alpha}_i)\bar{\alpha}_i. \end{aligned} \qquad (3.7)$$

In this chapter, we make the following assumptions:

Assumption 3.1 *The parameter uncertainty ΔA and unknown nonlinear function $f(x(k))$ are bounded in the Euclidean norm.*

Assumption 3.2 *The variables $\tau_i(k)$ $(i = 1, 2, \cdots, q)$ are time-varying and satisfy $d_m \le \tau_i(k) \le d_M$ where d_m and d_M are constant positive scalars representing the lower and upper bounds on the communication delays, respectively.*

3.2 Design of SMC

In this section, a switching function will first be presented for the uncertain nonlinear networked system (3.1) with stochastic communication delays. Then, a theorem will be given in order to determine the parameter appearing in the proposed switching function to satisfy the robustly mean square exponential stability in the sliding motion. We will show that the controller design problem in the sliding motion can be solved if two LMIs with an equality constraint are feasible.

Switching Surface

To begin with, we choose the following switching function:

$$s(k) = Gx(k) - GAx(k-1) \tag{3.8}$$

where G is designed so that GB is nonsingular and $G\widehat{D} = 0$, where $\widehat{D} \triangleq \begin{bmatrix} A_d & D \end{bmatrix}$. In this chapter, we select $G = B^{\mathrm{T}} P$ with $P > 0$ being a positive definite matrix to confirm the non-singularity of GB.

The ideal quasi-sliding mode satisfies

$$s(k+1) = s(k) = 0. \tag{3.9}$$

Solving the above for $u(k)$, the equivalent control law of the sliding motion is given by

$$u_{eq}(k) = -(GB)^{-1}G\Delta Ax(k) - f(x(k)). \tag{3.10}$$

Substituting (3.10) as $u(k)$ into (3.1) yields

$$\begin{cases} x(k+1) = A_K x(k) + A_d \tilde{x}(k) + Dg(x(k), \tilde{x}(k), k)\omega(k) \\ \tilde{x}(k) = \displaystyle\sum_{i=1}^{q} \alpha_i(k)x(k-\tau_i(k)) \end{cases} \tag{3.11}$$

where $A_K \triangleq A + \Delta A - B(GB)^{-1}G\Delta A$. The expression (3.11) is the sliding mode dynamics of system (3.1) in the specified switching surface $s(k+1) = s(k) = 0$.

Before stating the designing goal, we introduce the following stability concept for the system (3.11).

Definition 3.1 *The system (3.11) is said to be robustly exponentially stable in the mean square if there exist constants $\zeta > 0$ and $0 < \epsilon < 1$ such that every solution of (3.11) satisfies*

$$\mathbb{E}\{\|x(k)\|^2\} \leq \zeta \epsilon^k \sup_{-d_M \leq i \leq 0} \mathbb{E}\{\|x(i)\|^2\}, \qquad \forall k \geq 0, \tag{3.12}$$

for all admissible parameter uncertainties.

In this chapter, it is our objective to synthesize an SMC law such that the state trajectory in (3.1) is globally driven onto the specified sliding surface, leading to a non-increasing zigzag motion along the sliding surface in subsequent time and, at the same time, the system dynamics on the sliding surface are guaranteed to be robustly exponentially stable in the mean square.

Performances of the Sliding Motion

First of all, we introduce the following lemmas which will be used in this chapter.

Lemma 3.1 *For any real vectors a, b and matrix $P > 0$ of compatible dimensions,*

$$a^{\mathrm{T}}b + b^{\mathrm{T}}a \le a^{\mathrm{T}}Pa + b^{\mathrm{T}}P^{-1}b. \tag{3.13}$$

Lemma 3.2 *(Schur Complement Lemma) Given constant matrices $\mathcal{S}_1, \mathcal{S}_2, \mathcal{S}_3$ where $\mathcal{S}_1 = \mathcal{S}_1^{\mathrm{T}}$ and $0 < \mathcal{S}_2 = \mathcal{S}_2^{\mathrm{T}}$, then $\mathcal{S}_1 + \mathcal{S}_3^{\mathrm{T}}\mathcal{S}_2^{-1}\mathcal{S}_3 < 0$ if and only if*

$$\begin{bmatrix} \mathcal{S}_1 & \mathcal{S}_3^{\mathrm{T}} \\ \mathcal{S}_3 & -\mathcal{S}_2 \end{bmatrix} < 0 \quad \text{or} \quad \begin{bmatrix} -\mathcal{S}_2 & \mathcal{S}_3 \\ \mathcal{S}_3^{\mathrm{T}} & \mathcal{S}_1 \end{bmatrix} < 0. \tag{3.14}$$

Lemma 3.3 *(S-procedure) Let $N = N^{\mathrm{T}}$, H and E be real matrices of appropriate dimensions, and F satisfy (3.5). Then $N + HFE + E^{\mathrm{T}}F^{\mathrm{T}}H^{\mathrm{T}} < 0$ if and only if there exists a positive scalar ε such that $N + \varepsilon HH^{\mathrm{T}} + \varepsilon^{-1}E^{\mathrm{T}}E < 0$ or, equivalently,*

$$\begin{bmatrix} N & \varepsilon H & E^{\mathrm{T}} \\ \varepsilon H^{\mathrm{T}} & -\varepsilon I & 0 \\ E & 0 & -\varepsilon I \end{bmatrix} < 0. \tag{3.15}$$

The following theorem gives the sufficient condition in terms of LMIs with equality constraints for the global mean square asymptotic stability of system (3.11) in the sliding motion.

Theorem 3.1 *The system (3.11) is robustly exponentially stable in the mean square if there exist positive definite matrices $P > 0$, $Q_j > 0$ $(j = 1, 2, \cdots, q)$, and positive scalars $\varepsilon > 0$, $\lambda^* > 0$ satisfying*

$$\begin{bmatrix} \Omega_{11} & 0 & A^{\mathrm{T}}P & 0 & A^{\mathrm{T}}P & 0 & 0 & \varepsilon E^{\mathrm{T}} \\ * & \Omega_{22} & \widehat{Z}^{\mathrm{T}}A_d^{\mathrm{T}}P & \lambda^*\widehat{Z}^{\mathrm{T}} & 0 & 0 & 0 & 0 \\ * & * & -P & 0 & 0 & 0 & PH & 0 \\ * & * & * & -\frac{1}{\rho_2}\lambda^*I & 0 & 0 & 0 & 0 \\ * & * & * & * & -P & 0 & PH & 0 \\ * & * & * & * & * & -B^{\mathrm{T}}PB & \sqrt{3}B^{\mathrm{T}}PH & 0 \\ * & * & * & * & * & * & -\varepsilon I & 0 \\ * & * & * & * & * & * & * & -\varepsilon I \end{bmatrix} < 0, \tag{3.16}$$

$$D^{\mathrm{T}}PD < \lambda^*I, \tag{3.17}$$

$$B^{\mathrm{T}}P\widehat{D} = 0, \tag{3.18}$$

where

$$
\begin{aligned}
\Omega_{11} &= \lambda^* \rho_1 I - P + \sum_{j=1}^{q} (d_M - d_m + 1) Q_j, \\
\Omega_{22} &= \mathrm{diag}\{-Q_1 + \widetilde{A}_1, -Q_2 + \widetilde{A}_2, \cdots, -Q_q + \widetilde{A}_q\}, \\
\widetilde{A}_j &= \bar{\alpha}_j (1 - \bar{\alpha}_j)(A_d^{\mathrm{T}} P A_d + \rho_2 \lambda^* I), \quad j = 1, 2, \cdots, q. \\
\widehat{Z} &= \begin{bmatrix} \bar{\alpha}_1 I & \bar{\alpha}_2 I & \cdots & \bar{\alpha}_q I \end{bmatrix}.
\end{aligned}
$$

Proof *Let* $\Theta_j(k) \triangleq \{x(k - \tau_j(k)), x(k - \tau_j(k) + 1), \cdots, x(k)\}$ $(j = 1, 2, \cdots, q)$ *and define*

$$
\mathscr{X}(k) = \left\{ \Theta_1(k) \bigcup \Theta_2(k) \bigcup \cdots \bigcup \Theta_n(k) \right\} = \bigcup_{j=1}^{q} \Theta_j(k).
$$

Choose the following Lyapunov functional for system (3.11):

$$
W(\mathscr{X}(k)) = \sum_{i=1}^{3} W_i(k) \tag{3.19}
$$

where

$$
\begin{aligned}
W_1(k) &= x^{\mathrm{T}}(k) P x(k), \\
W_2(k) &= \sum_{j=1}^{q} \sum_{i=k-\tau_j(k)}^{k-1} x^{\mathrm{T}}(i) Q_j x(i), \\
W_3(k) &= \sum_{j=1}^{q} \sum_{m=-d_M+1}^{-d_m} \sum_{i=k+m}^{k-1} x^{\mathrm{T}}(i) Q_j x(i),
\end{aligned}
$$

with $P > 0$, $Q_j > 0$ $(j = 1, 2, \cdots, q)$ *being matrices to be determined. Then along the trajectory of system* (3.11), *we have*

$$
\begin{aligned}
\mathbb{E}\{\Delta W | \mathscr{X}(k)\} &\triangleq \mathbb{E}\{W(\mathscr{X}(k+1)) | \mathscr{X}(k)\} - W(\mathscr{X}(k)) \\
&= \mathbb{E}\{(W(\mathscr{X}(k+1)) - W(\mathscr{X}(k))) | \mathscr{X}(k)\} \\
&= \sum_{i=1}^{3} \mathbb{E}\{\Delta W_i | \mathscr{X}(k)\}.
\end{aligned} \tag{3.20}
$$

From (3.11), *we can obtain that*

$$
\begin{aligned}
\mathbb{E}\{\Delta W_1 | \mathscr{X}(k)\} &= \mathbb{E}\left\{ \left(x^{\mathrm{T}}(k+1) P x(k+1) - x^{\mathrm{T}}(k) P x(k) \right) \middle| \mathscr{X}(k) \right\} \\
&= \mathbb{E}\Big\{ x^{\mathrm{T}}(k)(A_K^{\mathrm{T}} P A_K - P) x(k) + \tilde{x}^{\mathrm{T}}(k) A_d^{\mathrm{T}} P A_d \tilde{x}(k) \\
&\quad + g^{\mathrm{T}}(x(k), \tilde{x}(k), k) D^{\mathrm{T}} P D g(x(k), \tilde{x}(k), k) \\
&\quad + 2 x^{\mathrm{T}}(k) A_K^{\mathrm{T}} P A_d \tilde{x}(k) \Big| \mathscr{X}(k) \Big\}.
\end{aligned} \tag{3.21}
$$

Denoting $\bar{A} = -B(GB)^{-1}G\Delta A$, by Lemma 3.1, we have

$$A_K^{\mathrm{T}} P A_K \leq 2(A + \Delta A)^{\mathrm{T}} P (A + \Delta A) + 2\bar{A}^{\mathrm{T}} P \bar{A} \tag{3.22}$$

and

$$
\begin{aligned}
2x^{\mathrm{T}}(k) A_K^{\mathrm{T}} P A_d \tilde{x}(k) =& 2x^{\mathrm{T}}(k)(A + \Delta A + \bar{A})^{\mathrm{T}} P A_d \tilde{x}(k) \\
\leq & 2x^{\mathrm{T}}(k)(A + \Delta A)^{\mathrm{T}} P A_d \tilde{x}(k) \\
& + x^{\mathrm{T}}(k)\bar{A}^{\mathrm{T}} P \bar{A} x(k) + \tilde{x}^{\mathrm{T}}(k) A_d^{\mathrm{T}} P A_d \tilde{x}(k).
\end{aligned}
\tag{3.23}
$$

It is noticed from (3.3) and (3.17) that

$$
\begin{aligned}
& g^{\mathrm{T}}(x(k), \tilde{x}(k), k) D^{\mathrm{T}} P D g(x(k), \tilde{x}(k), k) \\
\leq & \lambda_{\max}(D^{\mathrm{T}} P D) g^{\mathrm{T}}(x(k), \tilde{x}(k), k) g(x(k), \tilde{x}(k), k) \\
\leq & \lambda^* \Big(\rho_1 x^{\mathrm{T}}(k) x(k) + \rho_2 \tilde{x}^{\mathrm{T}}(k) \tilde{x}(k) \Big).
\end{aligned}
\tag{3.24}
$$

Introducing new variables by

$$\tilde{\alpha}_i(k) = \alpha_i(k) - \bar{\alpha}_i, i = 1, 2, \cdots, q, \tag{3.25}$$

then, the mean and variance of $\tilde{\alpha}_i(k)$ can be easily obtained by

$$\mathbb{E}\{\tilde{\alpha}_i(k)\} = 0, \qquad \sigma_{\tilde{\alpha}_i}^2 = \bar{\alpha}_i(1 - \bar{\alpha}_i). \tag{3.26}$$

Therefore,

$$
\begin{aligned}
& \mathbb{E}\left\{ \tilde{x}^{\mathrm{T}}(k) A_d^{\mathrm{T}} P A_d \tilde{x}(k) \Big| \mathscr{X}(k) \right\} \\
=& \mathbb{E}\bigg\{ \Big(\sum_{i=1}^{q} (\bar{\alpha}_i + \tilde{\alpha}_i(k)) x(k - \tau_i(k)) \Big)^{\mathrm{T}} A_d^{\mathrm{T}} P \\
& \times A_d \Big(\sum_{i=1}^{q} (\bar{\alpha}_i + \tilde{\alpha}_i(k)) x(k - \tau_i(k)) \Big) \Big| \mathscr{X}(k) \bigg\} \\
=& \Big(\sum_{i=1}^{q} \bar{\alpha}_i x(k - \tau_i(k)) \Big)^{\mathrm{T}} A_d^{\mathrm{T}} P A_d \Big(\sum_{i=1}^{q} \bar{\alpha}_i x(k - \tau_i(k)) \Big) \\
& + \sum_{i=1}^{q} \sigma_{\tilde{\alpha}_i}^2 x^{\mathrm{T}}(k - \tau_i(k)) A_d^{\mathrm{T}} P A_d x(k - \tau_i(k))
\end{aligned}
\tag{3.27}
$$

and

$$
\begin{aligned}
& \mathbb{E}\left\{ x^{\mathrm{T}}(k)(A + \Delta A)^{\mathrm{T}} P A_d \tilde{x}(k) \Big| \mathscr{X}(k) \right\} \\
=& x^{\mathrm{T}}(k)(A + \Delta A)^{\mathrm{T}} P A_d \Big(\sum_{i=1}^{q} \bar{\alpha}_i x(k - \tau_i(k)) \Big).
\end{aligned}
\tag{3.28}
$$

Taking (3.21)-(3.28) into consideration and noting that $G = B^{\mathrm{T}}P$, we have

$$\mathbb{E}\{\Delta W_1 | \mathscr{X}(k)\}$$

$$\leq x^{\mathrm{T}}(k)\Big(2(A + \Delta A)^{\mathrm{T}}P(A + \Delta A)$$

$$+ 3\Delta A^{\mathrm{T}}PB(B^{\mathrm{T}}PB)^{-1}B^{\mathrm{T}}P\Delta A + \lambda^*\rho_1 I - P\Big)x(k)$$

$$+ \Big(\sum_{i=1}^{q}\bar{\alpha}_i x(k - \tau_i(k))\Big)^{\mathrm{T}}(A_d^{\mathrm{T}}PA_d + \lambda^*\rho_2 I)\Big(\sum_{i=1}^{q}\bar{\alpha}_i x(k - \tau_i(k))\Big) \quad (3.29)$$

$$+ \sum_{i=1}^{q}\sigma_{\bar{\alpha}_i}^2 x^{\mathrm{T}}(k - \tau_i(k))(A_d^{\mathrm{T}}PA_d + \lambda^*\rho_2 I)x(k - \tau_i(k))$$

$$+ 2x^{\mathrm{T}}(k)(A + \Delta A)^{\mathrm{T}}PA_d\Big(\sum_{i=1}^{q}\bar{\alpha}_i x(k - \tau_i(k))\Big).$$

Next, it can be derived that

$$\mathbb{E}\{\Delta W_2 | \mathscr{X}(k)\}$$

$$\leq \mathbb{E}\Bigg\{\sum_{j=1}^{q}\Big(x^{\mathrm{T}}(k)Q_j x(k) - x^{\mathrm{T}}(k - \tau_j(k))Q_j x(k - \tau_j(k))$$

$$+ \sum_{i=k-d_M+1}^{k-d_m} x^{\mathrm{T}}(i)Q_j x(i)\Big)\Bigg| \mathscr{X}(k)\Bigg\} \quad (3.30)$$

and

$$\mathbb{E}\{\Delta W_3 | \mathscr{X}(k)\} = \mathbb{E}\Bigg\{\sum_{j=1}^{q}\Big((d_M - d_m)x^{\mathrm{T}}(k)Q_j x(k)$$

$$- \sum_{i=k-d_M+1}^{k-d_m} x^{\mathrm{T}}(i)Q_j x(i)\Big)\Bigg| \mathscr{X}(k)\Bigg\}. \quad (3.31)$$

Letting

$$\xi(k) = \begin{bmatrix} x(k) \\ x(k - \tau_1(k)) \\ x(k - \tau_2(k)) \\ \vdots \\ x(k - \tau_q(k)) \end{bmatrix},$$

the combination of (3.29)-(3.31) results in

$$\mathbb{E}\{\Delta W | \mathscr{X}(k)\} \leq \xi^{\mathrm{T}}(k)\Upsilon\xi(k) \quad (3.32)$$

where

$$\Upsilon \triangleq \begin{bmatrix} \Upsilon_{11} & \Upsilon_{12} \\ \Upsilon_{12}^{\mathrm{T}} & \Upsilon_{22} \end{bmatrix},$$

$$\Upsilon_{11} = \Omega_{11} + 2(A + \Delta A)^{\mathrm{T}} P(A + \Delta A) + 3\Delta A^{\mathrm{T}} PB(B^{\mathrm{T}} PB)^{-1} B^{\mathrm{T}} P\Delta A,$$

$$\Upsilon_{12} = \Omega_{22} + \widehat{Z}^{\mathrm{T}} A_d^{\mathrm{T}} P A_d \widehat{Z} + \rho_2 \lambda^* \widehat{Z}^{\mathrm{T}} \widehat{Z},$$

$$\Upsilon_{22} = (A + \Delta A)^{\mathrm{T}} P A_d \widehat{Z}.$$

By the Schur Complement, it is easy to see that $\Upsilon < 0$ *is equivalent to*

$$\Xi \triangleq \begin{bmatrix} \Omega_{11} & 0 & (A + \Delta A)^{\mathrm{T}} P & 0 & (A + \Delta A)^{\mathrm{T}} P & \sqrt{3}\Delta A^{\mathrm{T}} PB \\ * & \Omega_{22} & \widehat{Z}^{\mathrm{T}} A_d^{\mathrm{T}} P & \lambda^* \widehat{Z}^{\mathrm{T}} & 0 & 0 \\ * & * & -P & 0 & 0 & 0 \\ * & * & * & -\frac{1}{\rho_2}\lambda^* I & 0 & 0 \\ * & * & * & * & -P & 0 \\ * & * & * & * & * & -B^{\mathrm{T}} PB \end{bmatrix} < 0. \tag{3.33}$$

In order to eliminate the parameter uncertainty ΔA *in (3.33), we rewrite* Ξ *as follows*

$$\Xi = \bar{\Xi} + \widehat{H} F \widehat{E} + \widehat{E}^{\mathrm{T}} F \widehat{H}^{\mathrm{T}} \tag{3.34}$$

where

$$\bar{\Xi} = \begin{bmatrix} \Omega_{11} & 0 & A^{\mathrm{T}} P & 0 & A^{\mathrm{T}} P & 0 \\ * & \Omega_{22} & \widehat{Z}^{\mathrm{T}} A_d^{\mathrm{T}} P & \lambda^* \widehat{Z}^{\mathrm{T}} & 0 & 0 \\ * & * & -P & 0 & 0 & 0 \\ * & * & * & -\frac{1}{\rho_2}\lambda^* I & 0 & 0 \\ * & * & * & * & -P & 0 \\ * & * & * & * & * & -B^{\mathrm{T}} PB \end{bmatrix},$$

$$\widehat{H} = \begin{bmatrix} 0 & 0 & H^{\mathrm{T}} P & 0 & H^{\mathrm{T}} P & \sqrt{3} H^{\mathrm{T}} PB \end{bmatrix}^{\mathrm{T}},$$

$$\widehat{E} = \begin{bmatrix} E & 0 & 0 & 0 & 0 & 0 \end{bmatrix}.$$

By Lemma 3.3, we can easily know that LMI (3.16) is equivalent to $\Xi < 0$ *for some* $\varepsilon > 0$, *which means that* $\Upsilon < 0$ *can be achieved. Hence, for all* $\xi(k) \neq 0$, $\mathbb{E}\{\Delta W | \mathscr{X}(k)\} \leq \xi^{\mathrm{T}}(k) \Upsilon \xi(k) < 0$. *Therefore, the robustly exponential stability of system (3.11) can be confirmed in the mean square sense. The proof is complete.*

Computational Algorithm

Notice that the condition in Theorem 3.1 is given as the feasibility problem of two LMIs (3.16)-(3.17) with an equality constraint (3.18). Using the algorithm proposed in [167], as the condition $B^{\mathrm{T}}P\widehat{D} = 0$ can be expressed by $\mathrm{tr}[(B^{\mathrm{T}}P\widehat{D})^{\mathrm{T}}B^{\mathrm{T}}P\widehat{D}] = 0$, we first present an inequality $(B^{\mathrm{T}}P\widehat{D})^{\mathrm{T}}B^{\mathrm{T}}P\widehat{D} \leq \gamma I$ which, by the Schur Complement Lemma, is equivalent to

$$\begin{bmatrix} -\gamma I & \widehat{D}^{\mathrm{T}}PB \\ B^{\mathrm{T}}P\widehat{D} & -I \end{bmatrix} \leq 0. \tag{3.35}$$

Then, the original nonconvex feasibility problem can be converted into the following minimization problem:

$$\min \gamma \quad \text{subject to} \quad (3.16), (3.17) \text{ and } (3.35). \tag{3.36}$$

If this minimum equals zero, the solutions will satisfy the LMIs (3.16)-(3.17), and the equality $B^{\mathrm{T}}P\widehat{D} = 0$. Thus, the robustly mean square exponential stability of system (3.11) is achievable.

3.3   Sliding Mode Controller

In this section, by applying the definition of discrete-time sliding mode reaching condition $(DSMRC)$ for discrete-time systems in [212], we will synthesize a sliding mode controller to ensure the $DSMRC$ for system (3.1) with specified sliding surface (3.9). In other words, the trajectory of system (3.1) starting from any initial state will (i) be globally driven onto the sliding surface (3.9) in finite time; (ii) stay within a specified band once it has crossed the sliding surface the first time, and then cross the surface again in every sampling period, resulting in a successive and non-increasing zigzag motion along the sliding surface.

First, based on the reaching law proposed in [212], we can easily check that the designing goals (i) and (ii) are achieved if the following inequalities are satisfied:

$$\begin{cases} \Delta s_i(k) = s_i(k+1) - s_i(k) \leq -\kappa\mu_i \cdot \mathbf{sgn}[s_i(k)] - \kappa\varrho_i s_i(k) \\ \qquad \text{if } s_i(k) > 0 \\ \Delta s_i(k) = s_i(k+1) - s_i(k) \geq -\kappa\mu_i \cdot \mathbf{sgn}[s_i(k)] - \kappa\varrho_i s_i(k) \\ \qquad \text{if } s_i(k) < 0 \end{cases} \tag{3.37}$$

where κ represents the sampling period, $\mu_i > 0$ and $\varrho_i > 0$ are properly chosen scalars satisfying $0 < 1 - \kappa\varrho_i < 1$, $(i = 1, 2, \cdots, m)$.

Since the parameter uncertainty ΔA and the external disturbance $f(x(k))$

are both assumed to be bounded in the Euclidean norm, $\Delta_a \triangleq G\Delta Ax(k)$ and $\Delta_f \triangleq GBf(x(k))$ will also be bounded [212, 233]. Denote δ_a^i and δ_f^i as the ith element in Δ_a and Δ_f, respectively. Suppose the lower and upper bounds on Δ_a and Δ_f are known and given as follows:

$$
\begin{aligned}
\underline{\delta_a^i} \leq \delta_a^i \leq \overline{\delta_a^i}, \\
\underline{\delta_f^i} \leq \delta_f^i \leq \overline{\delta_f^i}, \qquad i = 1, 2, \cdots, m,
\end{aligned}
\tag{3.38}
$$

where $\underline{\delta_a^i}$, $\overline{\delta_a^i}$, $\underline{\delta_f^i}$ and $\overline{\delta_f^i}$ are all known constants. Then denote

$$
\begin{aligned}
\widehat{\Delta}_a &= \begin{bmatrix} \widehat{\delta}_a^1 & \widehat{\delta}_a^2 & \cdots & \widehat{\delta}_a^m \end{bmatrix}^{\mathrm{T}}, \\
\widehat{\delta}_a^i &= \frac{\overline{\delta_a^i} + \underline{\delta_a^i}}{2}, \\
\widetilde{\Delta}_a &= \mathrm{diag}\left\{ \widetilde{\delta}_a^1, \widetilde{\delta}_a^2, \cdots, \widetilde{\delta}_a^m \right\}, \\
\widetilde{\delta}_a^i &= \frac{\overline{\delta_a^i} - \underline{\delta_a^i}}{2}, \\
\widehat{\Delta}_f &= \begin{bmatrix} \widehat{\delta}_f^1 & \widehat{\delta}_f^2 & \cdots & \widehat{\delta}_f^m \end{bmatrix}^{\mathrm{T}}, \\
\widehat{\delta}_f^i &= \frac{\overline{\delta_f^i} + \underline{\delta_f^i}}{2}, \\
\widetilde{\Delta}_a &= \mathrm{diag}\left\{ \widetilde{\delta}_f^1, \widetilde{\delta}_f^2, \cdots, \widetilde{\delta}_f^m \right\}, \\
\widetilde{\delta}_f^i &= \frac{\overline{\delta_f^i} - \underline{\delta_f^i}}{2}, \qquad i = 1, 2, \cdots, m.
\end{aligned}
\tag{3.39}
$$

Now we are ready to give the design technique of the robust SMC controller.

Theorem 3.2 *Suppose Problem* (3.36) *is solvable. For the uncertain nonlinear networked system* (3.1) *with sliding surface* (3.9), *where* $G = B^{\mathrm{T}}P$ *and* P *is the solution to* (3.36), *if the SMC law is given as*

$$
\begin{aligned}
u(k) = &- (GB)^{-1}\Big(\kappa U \cdot \boldsymbol{sgn}[s(k)] + \kappa V s(k) + GAx(k-1) \\
&- Gx(k) + (\widehat{\Delta}_a + \widehat{\Delta}_f) + (\widetilde{\Delta}_a + \widetilde{\Delta}_f)\boldsymbol{sgn}[s(k)] \Big),
\end{aligned}
\tag{3.40}
$$

with

$$
\begin{aligned}
U &= \mathrm{diag}\left\{ \mu_1, \mu_2, \cdots, \mu_m \right\} \in \mathbb{R}^{m \times m}, \\
V &= \mathrm{diag}\left\{ \varrho_1, \varrho_2, \cdots, \varrho_m \right\} \in \mathbb{R}^{m \times m}, \qquad i = 1, 2, \cdots, m,
\end{aligned}
$$

then the discrete-time sliding mode reaching condition of system (3.1) *with specified sliding surface* (3.9) *is satisfied. In other words, design goals* (i) *and* (ii) *are achieved simultaneously.*

Proof *By* (3.40), *with the switching function defined in* (3.8), *we obtain*

$$\Delta s(k) = s(k+1) - s(k)$$
$$= G\Delta Ax(k) + GB\big(u(k) + f(x(k))\big) - Gx(k) + GAx(k-1)$$
$$= -\kappa U \cdot \boldsymbol{sgn}[s(k)] - \kappa V s(k) + G\Delta Ax(k)$$
$$- \Big(\widehat{\Delta}_a + \widetilde{\Delta}_a \boldsymbol{sgn}[s(k)]\Big) + GBf(x(k)) \tag{3.41}$$
$$- \Big(\widehat{\Delta}_f + \widetilde{\Delta}_f \boldsymbol{sgn}[s(k)]\Big).$$

Noting (3.39) *and* (3.40), *we can easily arrive at* (3.37). *Therefore, the discrete-time sliding mode reaching condition of* (3.1) *is satisfied. The proof is complete.*

Remark 3.1 *We should point out that it is not difficult to extend the present results to systems that include polytopic parameter uncertainties, systems with stochastic disturbances, or systems with more general switching communication delays, within the same established LMI framework. We could also use more up-to-date delay-dependent analysis approaches to reduce possible conservatism without difficulties. The reason why we discuss system* (3.1) *is to make our theory more understandable and to avoid unnecessarily complicated notations.*

3.4   An Illustrative Example

In this section, we present an illustrative example to demonstrate the effectiveness of the proposed algorithm.

Consider the following nonlinear networked system with communication delays:

$$
\begin{cases}
x(k+1) = \left(\begin{bmatrix} 0.15 & -0.25 & 0 \\ 0 & 0.18 & 0 \\ 0.03 & 0 & -0.05 \end{bmatrix} \right. \\
\quad + \begin{bmatrix} 0.01 \\ 0.02 \\ 0 \end{bmatrix} \sin(0.6k) \begin{bmatrix} 0 & 0.01 & 0 \end{bmatrix} \Bigg) x(k) \\
\quad + \begin{bmatrix} -0.10 & 0.02 \\ 0.07 & 0 \\ 0.02 & -0.20 \end{bmatrix} \big(u(k) + f(x(k))\big) \\
\quad + \displaystyle\sum_{i=1}^{2} \begin{bmatrix} 0.03 & 0 & -0.01 \\ 0.02 & 0.03 & 0 \\ 0.04 & 0.05 & -0.01 \end{bmatrix} \alpha_i(k)x(k-\tau_i(k)) \\
\quad + \begin{bmatrix} 0.025 & 0.01 & 0 \\ 0 & -0.03 & 0 \\ 0.04 & 0.035 & -0.01 \end{bmatrix} g(x(k), \tilde{x}(k), k)\omega(k), \\
x(k) = \varphi(k), \quad k = -d_M, -d_M+1, \cdots, 0.
\end{cases}
\tag{3.42}
$$

Let $f(x(k)) = 0.3\sin(x(k))$ and $g(x(k), \tilde{x}(k), k) = 0.5x(k) + 0.5\tilde{x}(k)$. In order to design the explicit SMC controller, we suppose $G\Delta Ax(k)$ and $GBf(x(k))$ are bounded by the following conditions:

$$
\begin{aligned}
\underline{\delta_a^j} &= -\|GH\|\|Ex(k)\|, \\
\overline{\delta_a^j} &= \|GH\|\|Ex(k)\|, \\
\underline{\delta_f^j} &= -0.3\|GB\sin(x(k))\|, \\
\overline{\delta_f^j} &= 0.3\|GB\sin(x(k))\|.
\end{aligned}
\tag{3.43}
$$

Assume that the time-varying communication delays satisfy $2 \leq \tau_i(k) \leq 6$ $(i = 1, 2)$, and

$$
\bar{\alpha}_1 = \mathbb{E}\{\alpha_1(k)\} = 0.8, \quad \bar{\alpha}_2 = \mathbb{E}\{\alpha_2(k)\} = 0.6.
$$

Then, we can obtain

$$
\sigma_{\bar{\alpha}_1}^2 = 0.16, \quad \sigma_{\bar{\alpha}_2}^2 = 0.24.
$$

Using the Matlab LMI toolbox to solve problem (3.36), we have

$$
P = \begin{bmatrix} 806501 & -0.0454 & -0.0401 \\ -0.0454 & 9.3349 & -0.0285 \\ -0.0401 & -0.0285 & 8.8057 \end{bmatrix},
$$

$$
Q1 = \begin{bmatrix} 0.6733 & 0.0485 & -0.0042 \\ 0.0485 & 0.5575 & -0.0028 \\ -0.0042 & -0.0028 & 0.6937 \end{bmatrix},
$$

$$
Q2 = \begin{bmatrix} 0.6007 & 0.0443 & -0.0025 \\ 0.0443 & 0.4840 & -0.0023 \\ -0.0025 & -0.0023 & 0.3157 \end{bmatrix},
$$

$$
G = \begin{bmatrix} -0.8690 & 0.6574 & 0.1781 \\ 0.1810 & 0.0048 & -1.7619 \end{bmatrix},
$$

and $\gamma = 3.707945 \times 10^{-12}$ (hence the constraint $G\hat{D} = 0$ is satisfied). Choose $\kappa = 0.05$ (and $\kappa = 0.005$ for comparison), $\mu_j = 1$ and $\varrho_j = 1$ $(j = 1, 2)$. Then, it follows from Theorem 3.2 that the desired SMC law can be expressed with all known parameters. The simulation results are shown in Fig. 3.1 to Fig. 3.10, which confirm that the desired requirements are well achieved.

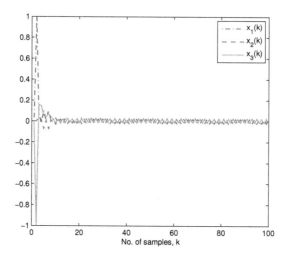

FIGURE 3.1: The trajectory of state $x(k)$ ($\kappa = 0.05$).

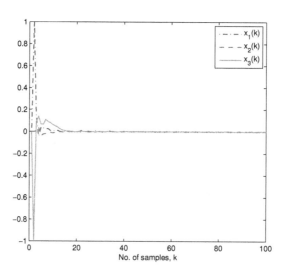

FIGURE 3.2: The trajectory of state $x(k)$ ($\kappa = 0.005$).

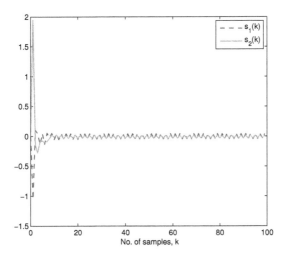

FIGURE 3.3: The trajectory of sliding variable $s(k)$ ($\kappa = 0.05$).

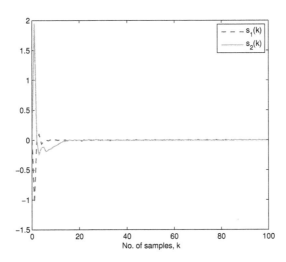

FIGURE 3.4: The trajectory of sliding variable $s(k)$ ($\kappa = 0.005$).

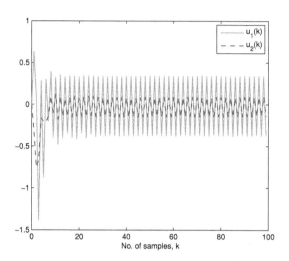

FIGURE 3.5: The control signals $u(k)$ ($\kappa = 0.05$).

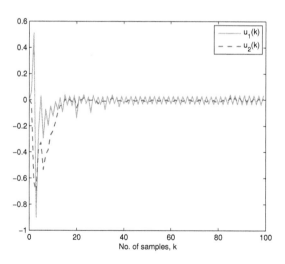

FIGURE 3.6: The control signals $u(k)$ ($\kappa = 0.005$).

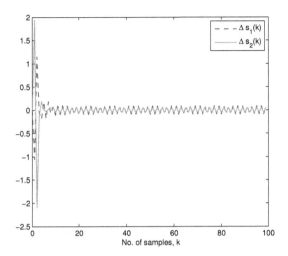

FIGURE 3.7: The control signals $u(k)$ ($\kappa = 0.05$).

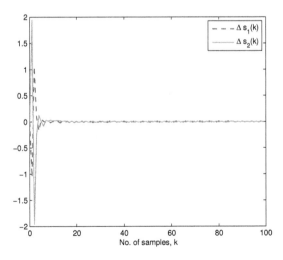

FIGURE 3.8: The control signals $u(k)$ ($\kappa = 0.005$).

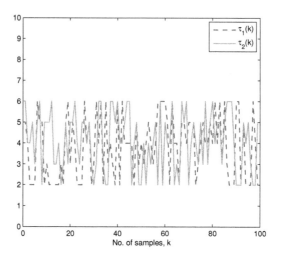

FIGURE 3.9: The time-varying delays $\tau_i(k)$ $(i = 1, 2)$.

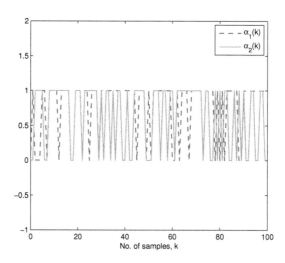

FIGURE 3.10: The Bernoulli sequences $\alpha_i(k)$ $(i = 1, 2)$.

3.5 Summary

A robust SMC design problem for a class of uncertain nonlinear networked systems with multiple stochastic communication delays has been studied. By means of LMIs, a sufficient condition for the robustly exponential stability of the system dynamics on the specified sliding surface has been derived. By the reaching condition proposed in this chapter, an SMC controller has been designed to globally drive the state trajectory onto the specified surface, which gives rise to a non-increasing zigzag motion along the surface. An illustrative numerical example has been given to show the applicability and effectiveness of the proposed method in this chapter.

4

Sliding Mode Control for Nonlinear
Networked Systems with Stochastic
Communication Delays

Rapid development in network technologies in the past few years has led to a revolution in engineering practices. Using a network as the basement of a feedback control loop is now widely applied in various types of engineering programs, which has gained a great deal of research interest. The usage of networks in control systems has many advantages such as low cost, reduced weight and simple installation, as well as some limitations including network-induced time-delays (also called communication delays) and data package loss caused mainly by complex working conditions and limited bandwidth. As a result, most of the literature on network systems has focused on how to eliminate or compensate for the effect caused by communication delays and data package loss. Many researchers have studied the stability and controller design problems for networked systems in the presence of deterministic communication delays. Recently, due to the fact that such kinds of time-delays usually appear in a random and time-varying fashion, the communication delays have been modeled in various *probabilistic* ways.

On the other hand, because of the advantage of strong robustness against model uncertainties, parameter variations and external disturbances, the sliding mode control (SMC) (also called variable structure control) problem of continuous-time systems has been extensively studied and widely applied in various fields. In recent years, the SMC problem for discrete-time systems has begun to receive increasing research attention simply because most control strategies are implemented in discrete time nowadays.

To date, the sliding mode control problem has not been studied for discrete-time uncertain nonlinear networked systems with or without stochastic time-varying delays. By using the discrete-time sliding motion concept, this chapter aims to design a state feedback controller such that (i) the system state trajectories are globally driven onto the pre-specified sliding surface with probability 1 which results in a non-increasing zigzag motion on the sliding surface; (ii) the globally mean square asymptotic stability of the system is guaranteed on the pre-specified sliding surface. The main contributions of this chapter are twofold: one is the new network-induced communication delay model that reflects the real world in a more realistic way, and the other is a new form of the discrete-time switching function which is proposed, for the first time, to deal with the sliding mode control problem.

The rest of this chapter is arranged as follows. Section 4.1 formulates an uncertain nonlinear networked system with stochastic time-varying delays. In Section 4.2, a novel switching function is first designed and then an LMI-based sufficient condition is given to obtain the parameters in the proposed switching function for ensuring the globally mean square asymptotic stability of the system dynamics on the sliding surface. An SMC law is synthesized in Section 4.3 to drive the state trajectories onto the specified surface with probability 1. In Section 4.4, an illustrative numerical example is provided to show the effectiveness and usefulness of the proposed approach. Section 4.5 gives our conclusions.

4.1   Problem Formulation

Consider the following class of uncertain nonlinear networked systems with communication delays:

$$
\begin{cases}
x(k+1) = (A + \Delta A)x(k) + A_d \sum_{i=1}^{n} M_i x(k - \alpha_i(k)\tau_i(k)) \\
\qquad\quad + B\big(u(k) + f(x(k))\big) \\
x(k) = \varphi(k), \quad k = -d_M, -d_M + 1, \cdots, 0
\end{cases}
\tag{4.1}
$$

where $x(k) \in \mathbb{R}^n$ is the state vector, $u(k) \in \mathbb{R}^m$ is the control input, A, A_d, B are known constant real-valued matrices with appropriate dimensions. M_i $(i = 1, 2, \cdots, n)$ has the form of

$$
M_i = \mathrm{diag}\{\underbrace{0, 0, \cdots, 0}_{i-1}, 1, 0, \cdots, 0\} \in \mathbb{R}^{n \times n}, \quad i = 1, 2, \cdots, n.
\tag{4.2}
$$

The real-valued matrix ΔA represents the norm-bounded parameter uncertainty formulated by

$$
\Delta A = HFE
\tag{4.3}
$$

where H and E are known real constant matrices which characterize how the deterministic uncertain parameter in F enters the nominal matrix A. The matrix, F, which could also be time-varying, is an unknown matrix function meeting

$$
F^{\mathrm{T}} F \leq I.
\tag{4.4}
$$

The parameter uncertainty ΔA is said to be admissible if both (4.3) and (4.4) are satisfied.

The stochastic variables $\alpha_i(k) \in \mathbb{R}$ $(i = 1, 2, \cdots, n)$ are mutually uncorrelated Bernoulli distributed white sequences taking values on 0 and 1 with

$$\text{Prob}\{\alpha_i(k) = 1\} = \mathbb{E}\{\alpha_i(k)\} \triangleq \bar{\alpha}_i \tag{4.5}$$

where $\bar{\alpha}_i$ $(i = 1, 2, \cdots, n)$ are known positive scalars. Therefore, we have

$$\text{Prob}\{\alpha_i(k) = 0\} = 1 - \bar{\alpha}_i,$$
$$\sigma_{\alpha_i}^2 \triangleq \mathbb{E}\{(\alpha_i(k) - \bar{\alpha}_i)^2\} = (1 - \bar{\alpha}_i)\bar{\alpha}_i. \tag{4.6}$$

In this chapter, we make the following assumptions:

Assumption 4.1 *The parameter uncertainty ΔA and the unknown nonlinear function $f(x(k))$ are bounded in the Euclidean norm.*

Assumption 4.2 *The variables $\tau_i(k)$ $(i = 1, 2, \cdots, n)$, which correspond to the n different communication delays, are assumed to be time-varying and satisfy $d_m \leq \tau_i(k) \leq d_M$ $(i = 1, 2, \cdots, n)$, where d_m and d_M are constant positive scalars representing the lower and upper bounds on the communication delays, respectively.*

Remark 4.1 *In Assumption 4.2, although the n time-varying communication delays are assumed to have the same lower and upper bounds, they could actually be different at the same sampling instant k, as demonstrated in Fig. 4.9 later. Such a description could be used to describe the phenomenon that the state variable should have its individual communication delay at the same sample time k, which is really the case in many practical applications. To our knowledge, such a description is new in the literature.*

Remark 4.2 *The Bernoulli sequences $\alpha_i(k)$ $(k = 1, 2, \cdots, n)$ are employed in this chapter to characterize the communication delays switching between 0 (the communication of the network is excellent, and there is no delay) and $\tau_i(k)$ (the communication of the network is not so good). It is worth mentioning that, using the method proposed later in this chapter, it is not technically difficult to extend our main results to the situation that the communication delays are switching within the set $\{\tilde{\tau}_1, \tilde{\tau}_2, \cdots, \tilde{\tau}_q\}$, where $\tilde{\tau}_j$ $(j = 1, 2, \cdots, q)$ are different and also could be time-varying communication delays.*

4.2 Design of SMC

In this section, a novel switching function is first presented for the uncertain nonlinear networked system (4.1) with stochastic communication delays. Then, a theorem is given in order to determine the parameter appearing in the proposed switching function to satisfy the globally mean square asymptotic stability in the sliding motion. It is shown that the controller design problem in the sliding motion can be solved if an LMI with an equality constraint is feasible.

Switching Surface

To begin with, we propose the following switching function

$$s(k) = Gx(k) - GAx(k-1) \tag{4.7}$$

where G is designed such that GB is nonsingular and $GA_d = 0$. In this chapter, we choose $G = B^{\mathrm{T}}P$ with $P > 0$ being a positive definite matrix to confirm the non-singularity of GB.

The ideal quasi-sliding mode satisfies

$$s(k+1) = s(k) = 0. \tag{4.8}$$

Solving for $u(k)$, the equivalent control law of the sliding motion is given by

$$u_{eq}(k) = -(GB)^{-1}G\Delta Ax(k) - f(x(k)). \tag{4.9}$$

Substituting (4.9) into (4.1) yields

$$
\begin{aligned}
x(k+1) =& (A + \Delta A - B(GB)^{-1}G\Delta A)x(k) \\
& + A_d \cdot \sum_{i=1}^{n} M_i x(k - \alpha_i(k)\tau_i(k)),
\end{aligned} \tag{4.10}
$$

which gives the sliding mode dynamics of system (4.1) in the specified switching surface $s(k+1) = s(k) = 0$.

For the convenience of derivation, we now present another formulation of equation (4.10). Introducing stochastic variables by $\beta_i(k) = 1 - \alpha_i(k)$, we have

$$
\begin{aligned}
\mathbb{E}\{\beta_i(k)\} &\triangleq \bar{\beta}_i = 1 - \bar{\alpha}_i, \\
\sigma_{\beta_i}^2 &\triangleq \mathbb{E}\{(\beta_i(k) - \bar{\beta}_i)^2\} = \bar{\alpha}_i(1 - \bar{\alpha}_i).
\end{aligned} \tag{4.11}
$$

Then, (4.10) can be reformulated as follows:

$$
\begin{aligned}
x(k+1) =& (A + \Delta A - B(GB)^{-1}G\Delta A)x(k) \\
& + A_d \sum_{i=1}^{n} M_i x(k - \alpha_i(k)\tau_i(k)) \\
=& (A + \Delta A - B(GB)^{-1}G\Delta A)x(k) \\
& + A_d \sum_{i=1}^{n} M_i \Big[\beta_i(k)x(k) + \alpha_i(k)x(k - \tau_i(k))\Big] \\
=& A_K x(k) + \sum_{i=1}^{n} \alpha_i(k)A_d M_i x(k - \tau_i(k))
\end{aligned} \tag{4.12}
$$

where $A_K = A + \Delta A + \sum_{i=1}^{n} \beta_i(k) A_d M_i - B(GB)^{-1} G \Delta A$.

Before giving the designing goal, we introduce the following stability concept for the system (4.12).

Definition 4.1 *The system (4.12) is said to be mean square stable if, for any $\epsilon > 0$, there is a $\delta(\epsilon) > 0$ such that $\mathbb{E}\{\|x(k)\|^2\} < \epsilon$ $(k > 0)$ when $\sup_{-d_M \leq s \leq 0} \mathbb{E}\{\|\varphi(k)\|^2\} < \delta(\epsilon)$. In addition, if $\lim_{k \to \infty} \mathbb{E}\{\|x(k)\|^2\} = 0$ for any initial conditions, then system (4.12) is said to be globally mean square asymptotically stable.*

In this chapter, it is our objective to design an SMC law such that the state trajectories in (4.1) are globally driven onto (with probability 1) the specified sliding surface, and subsequently cause a non-increasing zigzag motion along the sliding surface. At the same time, the system dynamics on the sliding surface are guaranteed to be globally mean square asymptotically stable.

First of all, we introduce the following lemmas which will be used in this chapter.

Lemma 4.1 *For any real vectors a, b and matrix $P > 0$ of compatible dimensions,*

$$a^{\mathrm{T}} b + b^{\mathrm{T}} a \leq a^{\mathrm{T}} P a + b^{\mathrm{T}} P^{-1} b. \tag{4.13}$$

Lemma 4.2 *(Schur Complement Lemma) Given constant matrices S_1, S_2, S_3 where $S_1 = S_1^{\mathrm{T}}$ and $0 < S_2 = S_2^{\mathrm{T}}$, then $S_1 + S_3^{\mathrm{T}} S_2^{-1} S_3 < 0$ if and only if*

$$\begin{bmatrix} S_1 & S_3^{\mathrm{T}} \\ S_3 & -S_2 \end{bmatrix} < 0 \quad \text{or} \quad \begin{bmatrix} -S_2 & S_3 \\ S_3^{\mathrm{T}} & S_1 \end{bmatrix} < 0. \tag{4.14}$$

Lemma 4.3 *(S-procedure) Let $N = N^{\mathrm{T}}$, H and E be real matrices of appropriate dimensions, and F satisfy (4.4). Then $N + HFE + E^{\mathrm{T}} F^{\mathrm{T}} H^{\mathrm{T}} < 0$ if and only if there exists a positive scalar ε such that $N + \varepsilon H H^{\mathrm{T}} + \varepsilon^{-1} E^{\mathrm{T}} E < 0$ or, equivalently,*

$$\begin{bmatrix} N & \varepsilon H & E^{\mathrm{T}} \\ \varepsilon H^{\mathrm{T}} & -\varepsilon I & 0 \\ E & 0 & -\varepsilon I \end{bmatrix} < 0. \tag{4.15}$$

The following theorem gives the sufficient condition in terms of LMI for the globally mean square asymptotic stability of system (4.12) in the sliding motion.

Theorem 4.1 *The system (4.12) is globally mean square asymptotically stable if there exist positive definite matrices $P > 0$, $Q_i > 0$ $(i = 1, 2, \cdots, n)$,* *and a positive scalar $\varepsilon > 0$ satisfying*

$$
\begin{bmatrix}
\Omega_{11} & \Omega_{12} & A^{\mathrm{T}}P & Z^{\mathrm{T}}P & \sqrt{3}A^{\mathrm{T}}P & \sqrt{3}Z^{\mathrm{T}}P \\
* & \Omega_{22} & \widehat{A}_d^{\mathrm{T}}P & \widehat{A}_d^{\mathrm{T}}P & 0 & 0 \\
* & * & -P & 0 & 0 & 0 \\
* & * & * & -P & 0 & 0 \\
* & * & * & * & -P & 0 \\
* & * & * & * & * & -P \\
* & * & * & * & * & * \\
* & * & * & * & * & * \\
* & * & * & * & * & *
\end{bmatrix}
$$

$$
\begin{bmatrix}
0 & 0 & \varepsilon E^{\mathrm{T}} \\
0 & 0 & 0 \\
0 & PH & 0 \\
0 & 0 & 0 \\
0 & \sqrt{3}PH & 0 \\
0 & 0 & 0 \\
-B^{\mathrm{T}}PB & \sqrt{3}B^{\mathrm{T}}PH & 0 \\
* & -\varepsilon I & 0 \\
* & * & -\varepsilon I
\end{bmatrix} < 0 \qquad (4.16)
$$

$$
B^{\mathrm{T}}PA_d = 0
$$

where

$$
Z = \sum_{i=1}^{n} \bar{\beta}_i A_d M_i,
$$

$$
\Omega_{11} = -P + \sum_{i=1}^{n}(d_M - d_m + 1)Q_i + \sum_{i=1}^{n} 4\sigma_{\bar{\beta}_i}^2 M_i^{\mathrm{T}} A_d^{\mathrm{T}} P A_d M_i,
$$

$$
\tilde{A}_{di} = \sigma_{\bar{\alpha}_i}^2 M_i^{\mathrm{T}} A_d^{\mathrm{T}} P A_d M_i, \quad i = 1, 2, \cdots, n
$$

$$
\Omega_{12} = \begin{bmatrix} -\tilde{A}_{d1} & -\tilde{A}_{d2} & \cdots & -\tilde{A}_{dn} \end{bmatrix},
$$

$$
\Omega_{22} = \mathrm{diag}\left\{ 2\tilde{A}_{d1} - Q_1, 2\tilde{A}_{d2} - Q_2, \cdots, 2\tilde{A}_{dn} - Q_n \right\},
$$

$$
\widehat{A}_d = \begin{bmatrix} \bar{\alpha}_1 A_d M_1 & \bar{\alpha}_2 A_d M_2 & \cdots & \bar{\alpha}_n A_d M_n \end{bmatrix}.
$$

Proof *Let $\Theta_i(k) \triangleq \{x(k - \tau_i(k)), x(k - \tau_i(k) + 1), \cdots, x(k)\}$ ($i = 1, 2, \cdots, n$) and define*

$$
\mathscr{X}(k) = \left\{ \Theta_1(k) \bigcup \Theta_2(k) \bigcup \cdots \bigcup \Theta_n(k) \right\} = \bigcup_{i=1}^{n} \Theta_i(k). \qquad (4.17)
$$

Choose the following Lyapunov function for system (4.12):

$$
W(\mathscr{X}(k)) = W_1 + W_2 + W_3 \qquad (4.18)
$$

where

$$
\begin{aligned}
W_1 &= x^\mathrm{T}(k)Px(k), \\
W_2 &= \sum_{j=1}^{n}\sum_{i=k-\tau_j(k)}^{k-1} x^\mathrm{T}(i)Q_j x(i), \\
W_3 &= \sum_{j=1}^{n}\sum_{m=-d_M+2}^{-d_m+1}\sum_{i=k+m-1}^{k-1} x^\mathrm{T}(i)Q_j x(i),
\end{aligned}
$$

with $P > 0$ and $Q_j > 0$ $(j = 1, 2, \cdots, n)$ being matrices to be determined. Then, along the trajectory of system (4.12), we define

$$
\begin{aligned}
\mathbb{E}\left\{\Delta W | \mathscr{X}(k)\right\} \triangleq &\,\mathbb{E}\left\{W(\mathscr{X}(k+1)) | \mathscr{X}(k)\right\} - W(\mathscr{X}(k)) \\
=&\,\mathbb{E}\left\{\big(W(\mathscr{X}(k+1)) - W(\mathscr{X}(k))\big) | \mathscr{X}(k)\right\} \\
=&\,\mathbb{E}\left\{\Delta W_1 | \mathscr{X}(k)\right\} + \mathbb{E}\left\{\Delta W_2 | \mathscr{X}(k)\right\} \\
&+ \mathbb{E}\left\{\Delta W_3 | \mathscr{X}(k)\right\}.
\end{aligned} \tag{4.19}
$$

Denoting

$$
\widehat{A}_K \triangleq A + \Delta A + \sum_{i=1}^{n} \beta_i(k)A_d M_i,
$$

$$
\bar{A} \triangleq -B(GB)^{-1}G\Delta A,
$$

we can obtain from (4.12) that

$$
\begin{aligned}
x^\mathrm{T}(k+1)Px(k+1) =&\, x^\mathrm{T}(k)A_K^\mathrm{T} P A_K x(k) \\
&+ 2x^\mathrm{T}(k)\widehat{A}_K^\mathrm{T} P\Big(\sum_{i=1}^{n}\alpha_i(k)A_d M_i x(k-\tau_i(k)) \Big) \\
&+ 2x^\mathrm{T}(k)\bar{A}^\mathrm{T} P\Big(\sum_{i=1}^{n}\alpha_i(k)A_d M_i x(k-\tau_i(k)) \Big) \\
&+ \Big(\sum_{i=1}^{n}\alpha_i(k)A_d M_i x(k-\tau_i(k)) \Big)^\mathrm{T} \\
&\times P\Big(\sum_{i=1}^{n}\alpha_i(k)A_d M_i x(k-\tau_i(k)) \Big).
\end{aligned} \tag{4.20}
$$

By Lemma 4.1, we have

$$
\begin{aligned}
A_K^\mathrm{T} P A_K =&\,(\widehat{A}_K + \bar{A})^\mathrm{T} P(\widehat{A}_K + \bar{A}) \\
\leq&\, 2\widehat{A}_K^\mathrm{T} P\widehat{A}_K + 2\bar{A}^\mathrm{T} P\bar{A} \\
=&\,\Big(A + \Delta A + \sum_{i=1}^{n}\beta_i(k)A_d M_i\Big)^\mathrm{T} 2P \\
&\times \Big(A + \Delta A + \sum_{i=1}^{n}\beta_i(k)A_d M_i\Big) + 2\bar{A}^\mathrm{T} P\bar{A}
\end{aligned}
$$

$$\leq \left(A + \Delta A\right)^{\mathrm{T}} 4P\left(A + \Delta A\right) + \left(\sum_{i=1}^{n} \beta_i(k) A_d M_i\right)^{\mathrm{T}}$$
$$\times 4P\left(\sum_{i=1}^{n} \beta_i(k) A_d M_i\right) + 2\bar{A}^{\mathrm{T}} P \bar{A} \tag{4.21}$$

and

$$2x^{\mathrm{T}}(k)\bar{A}^{\mathrm{T}} P\left(\sum_{i=1}^{n} \alpha_i(k) A_d M_i x(k - \tau_i(k))\right)$$
$$\leq \left(\sum_{i=1}^{n} \alpha_i(k) A_d M_i x(k - \tau_i(k))\right)^{\mathrm{T}} P\left(\sum_{i=1}^{n} \alpha_i(k) A_d M_i x(k - \tau_i(k))\right) \tag{4.22}$$
$$+ x^{\mathrm{T}}(k)\bar{A}^{\mathrm{T}} P \bar{A} x(k).$$

Introducing new variables by

$$\tilde{\alpha}_i(k) = \alpha_i(k) - \bar{\alpha}_i,$$
$$\tilde{\beta}_i(k) = \beta_i(k) - \bar{\beta}_i, \quad i = 1, 2, \cdots, n \tag{4.23}$$

it is easy to see that the means and variances of $\tilde{\alpha}_i(k)$ and $\tilde{\beta}_i(k)$ are given by

$$\mathbb{E}\left\{\tilde{\alpha}_i(k)\right\} = \mathbb{E}\left\{\tilde{\beta}_i(k)\right\} = 0,$$
$$\sigma^2_{\tilde{\alpha}_i} = \sigma^2_{\tilde{\beta}_i} = \bar{\alpha}_i(1 - \bar{\alpha}_i). \tag{4.24}$$

Therefore,

$$\mathbb{E}\left\{\left(\sum_{i=1}^{n} \beta_i(k) A_d M_i x(k)\right)^{\mathrm{T}} P\left(\sum_{i=1}^{n} \beta_i(k) A_d M_i x(k)\right) \Big| \mathscr{X}(k)\right\}$$
$$= \mathbb{E}\left\{\left(\sum_{i=1}^{n} (\bar{\beta}_i + \tilde{\beta}_i(k)) A_d M_i x(k)\right)^{\mathrm{T}} P \times \left(\sum_{i=1}^{n} (\bar{\beta}_i + \tilde{\beta}_i(k)) A_d M_i x(k)\right) \Big| \mathscr{X}(k)\right\}$$
$$= \left(\sum_{i=1}^{n} \bar{\beta}_i A_d M_i x(k)\right)^{\mathrm{T}} P\left(\sum_{i=1}^{n} \bar{\beta}_i A_d M_i x(k)\right) + \sum_{i=1}^{n} \sigma^2_{\tilde{\beta}_i} x^{\mathrm{T}}(k) M_i^{\mathrm{T}} A_d^{\mathrm{T}} P A_d M_i x(k), \tag{4.25}$$

$$\mathbb{E}\left\{\left(x^{\mathrm{T}}(k)\widehat{A}_K^{\mathrm{T}} P\left(\sum_{i=1}^{n} \alpha_i(k) A_d M_i x(k - \tau_i(k))\right)\right) \Big| \mathscr{X}(k)\right\}$$
$$= x^{\mathrm{T}}(k)\left(A + \Delta A + \sum_{i=1}^{n} \bar{\beta}_i A_d M_i\right)^{\mathrm{T}} P\left(\sum_{i=1}^{n} \bar{\alpha}_i A_d M_i x(k - \tau_i(k))\right) \tag{4.26}$$
$$- x^{\mathrm{T}}(k) \sum_{i=1}^{n} \sigma^2_{\tilde{\alpha}_i} M_i^{\mathrm{T}} A_d^{\mathrm{T}} P A_d M_i x(k - \tau_i(k))$$

and

$$
\mathbb{E}\left\{\left(\sum_{i=1}^{n}\alpha_i(k)A_dM_ix(k-\tau_i(k))\right)^{\mathrm{T}}P\right.
$$

$$
\left.\times\left(\sum_{i=1}^{n}\alpha_i(k)A_dM_ix(k-\tau_i(k))\right)\middle|\mathscr{X}(k)\right\}
$$

$$
=\left(\sum_{i=1}^{n}\bar{\alpha}_iA_dM_ix(k-\tau_i(k))\right)^{\mathrm{T}}P\left(\sum_{i=1}^{n}\bar{\alpha}_iA_dM_ix(k-\tau_i(k))\right)
$$

$$
+\sum_{i=1}^{n}\sigma^2_{\bar{\alpha}_i}\left(A_dM_ix(k-\tau_i(k))\right)^{\mathrm{T}}P\left(A_dM_ix(k-\tau_i(k))\right).
$$

(4.27)

Since $G = B^{\mathrm{T}}P$, we have

$$
\bar{A}^{\mathrm{T}}P\bar{A} = \Delta A^{\mathrm{T}}PB(B^{\mathrm{T}}PB)^{-1}B^{\mathrm{T}}P\Delta A. \tag{4.28}
$$

Taking (4.20)-(4.28) into consideration, we can obtain

$$
\mathbb{E}\left\{\Delta W_1|\mathscr{X}(k)\right\}
$$

$$
=\mathbb{E}\left\{\left(x^{\mathrm{T}}(k+1)Px(k+1)-x^{\mathrm{T}}(k)Px(k)\right)\middle|\mathscr{X}(k)\right\}
$$

$$
\leq x^{\mathrm{T}}(k)\left[(A+\Delta A)^{\mathrm{T}}4P(A+\Delta A)\right.
$$

$$
+\left(\sum_{i=1}^{n}\bar{\beta}_iA_dM_i\right)^{\mathrm{T}}4P\left(\sum_{i=1}^{n}\bar{\beta}_iA_dM_i\right)+\sum_{i=1}^{n}4\sigma^2_{\bar{\beta}_i}M_i^{\mathrm{T}}A_d^{\mathrm{T}}PA_dM_i
$$

$$
\left.+3\Delta A^{\mathrm{T}}PB(B^{\mathrm{T}}PB)^{-1}B^{\mathrm{T}}P\Delta A-P\right]x(k)
$$

$$
+2x^{\mathrm{T}}(k)\left(A+\Delta A+\sum_{i=1}^{n}\bar{\beta}_iA_dM_i\right)^{\mathrm{T}}P\left(\sum_{i=1}^{n}\bar{\alpha}_iA_dM_ix(k-\tau_i(k))\right)
$$

$$
-2x^{\mathrm{T}}(k)\sum_{i=1}^{n}\sigma^2_{\bar{\alpha}_i}M_i^{\mathrm{T}}A_d^{\mathrm{T}}PA_dM_ix(k-\tau_i(k))
$$

$$
+\left(\sum_{i=1}^{n}\bar{\alpha}_iA_dM_ix(k-\tau_i(k))\right)^{\mathrm{T}}2P\left(\sum_{i=1}^{n}\bar{\alpha}_iA_dM_ix(k-\tau_i(k))\right)
$$

$$
+\sum_{i=1}^{n}\sigma^2_{\bar{\alpha}_i}\left(A_dM_ix(k-\tau_i(k))\right)^{\mathrm{T}}2P\left(A_dM_ix(k-\tau_i(k))\right).
$$

(4.29)

Next, it can be seen that

$$
\mathbb{E}\left\{\Delta W_2 \,|\, \mathscr{X}(k)\right\} = \mathbb{E}\left\{\sum_{j=1}^{n}\left(x^{\mathrm{T}}(k)Q_j x(k) - x^{\mathrm{T}}(k-\tau_j(k))Q_j x(k-\tau_j(k))\right.\right.
$$

$$
+ \sum_{i=k-\tau_j(k+1)+1}^{k-1} x^{\mathrm{T}}(i)Q_j x(i) - \sum_{i=k-\tau_j(k)+1}^{k-1} x^{\mathrm{T}}(i)Q_j x(i)\Bigg)\Bigg|\,\mathscr{X}(k)\Bigg\}
$$

$$
\leq \mathbb{E}\left\{\sum_{j=1}^{n}\left(x^{\mathrm{T}}(k)Q_j x(k) - x^{\mathrm{T}}(k-\tau_j(k))Q_j x(k-\tau_j(k))\right.\right.
$$

$$
+ \sum_{i=k-d_M+1}^{k-d_m} x^{\mathrm{T}}(i)Q_j x(i)\Bigg)\Bigg|\,\mathscr{X}(k)\Bigg\}.
$$

(4.30)

$$
\mathbb{E}\left\{\Delta W_3 \,|\, \mathscr{X}(k)\right\} = \mathbb{E}\left\{\sum_{j=1}^{n}\left((d_M - d_m)x^{\mathrm{T}}(k)Q_j x(k)\right.\right.
$$

(4.31)

$$
- \sum_{i=k-d_M+1}^{k-d_m} x^{\mathrm{T}}(i)Q_j x(i)\Bigg)\Bigg|\,\mathscr{X}(k)\Bigg\}.
$$

Let

$$
\eta(k) = \begin{bmatrix} x(k) \\ x^{\mathrm{T}}(k-\tau_1(k)) \\ x^{\mathrm{T}}(k-\tau_2(k)) \\ \vdots \\ x^{\mathrm{T}}(k-\tau_n(k)) \end{bmatrix},
$$

then

$$
\mathbb{E}\left\{\Delta W \,|\, \mathscr{X}(k)\right\} \leq \eta^{\mathrm{T}}(k)\Upsilon\eta(k)
$$

(4.32)

where

$$
\Upsilon = \begin{bmatrix}
\Omega_{11} & \Omega_{12} & (A+\Delta A)^{\mathrm{T}}P & Z^{\mathrm{T}}P & \sqrt{3}(A+\Delta A)^{\mathrm{T}}P & \sqrt{3}Z^{\mathrm{T}}P & \sqrt{3}\Delta A^{\mathrm{T}}PB \\
* & \Omega_{22} & \widehat{A}_d^{\mathrm{T}}P & \widehat{A}_d^{\mathrm{T}}P & 0 & 0 & 0 \\
* & * & -P & 0 & 0 & 0 & 0 \\
* & * & * & -P & 0 & 0 & 0 \\
* & * & * & * & -P & -P & 0 \\
* & * & * & * & * & -P & 0 \\
* & * & * & * & * & * & -B^{\mathrm{T}}PB
\end{bmatrix}.
$$

Rewrite Υ into the following form

$$\Upsilon = \begin{bmatrix} \Omega_{11} & \Omega_{12} & A^{\mathrm{T}}P & Z^{\mathrm{T}}P & \sqrt{3}A^{\mathrm{T}}P & \sqrt{3}Z^{\mathrm{T}}P & 0 \\ * & \Omega_{22} & \widehat{A}_d^{\mathrm{T}}P & \widehat{A}_d^{\mathrm{T}}P & 0 & 0 & 0 \\ * & * & -P & 0 & 0 & 0 & 0 \\ * & * & * & -P & 0 & 0 & 0 \\ * & * & * & * & -P & 0 & 0 \\ * & * & * & * & * & -P & 0 \\ * & * & * & * & * & * & -B^{\mathrm{T}}PB \end{bmatrix} \tag{4.33}$$

$$+ \widehat{H}F\widehat{E} + \widehat{E}^{\mathrm{T}}F\widehat{H}^{\mathrm{T}}$$

where

$$\widehat{H} = \begin{bmatrix} 0 & 0 & H^{\mathrm{T}}P & 0 & \sqrt{3}H^{\mathrm{T}}P & 0 & \sqrt{3}H^{\mathrm{T}}PB \end{bmatrix}^{\mathrm{T}},$$

$$\widehat{E} = \begin{bmatrix} E & 0 & 0 & 0 & 0 & 0 & 0 \end{bmatrix}.$$

Using Lemma 4.3, we can see that (4.16) is equivalent to $\Upsilon < 0$. Therefore, for all nonzero $\eta(k)$, $\mathbb{E}\{\Delta W | \mathscr{X}(k)\} \leq \eta^{\mathrm{T}}(k)\Upsilon\eta(k) < 0$. As a result, we can confirm the system (4.12) is globally mean square asymptotically stable. The proof is complete.

Computational Algorithm

It is seen that the condition in Theorem 4.1 is presented as the feasibility problem of an LMI with an equality constraint. Using the algorithm proposed in [167], as the condition $B^{\mathrm{T}}PA_d = 0$ is equivalent to $\mathbf{tr}[(B^{\mathrm{T}}PA_d)^{\mathrm{T}}B^{\mathrm{T}}PA_d] = 0$, we first introduce the condition $(B^{\mathrm{T}}PA_d)^{\mathrm{T}}B^{\mathrm{T}}PA_d \leq \gamma I$. By the Schur-complement, the condition can be expressed as

$$\begin{bmatrix} -\gamma I & A_d^{\mathrm{T}}PB \\ B^{\mathrm{T}}PA_d & -I \end{bmatrix} \leq 0. \tag{4.34}$$

Hence, the original nonconvex feasibility problem can be converted into the following minimization problem:

$$\min \gamma \quad \text{subject to} \quad (4.16) \text{ and } (4.34). \tag{4.35}$$

If this infinum equals zero, the solutions will satisfy the LMI (4.16) with the equality $B^{\mathrm{T}}PA_d = 0$. Thus, the globally mean square asymptotic stability of system (4.12) is achieved.

4.3 Sliding Mode Controller

In this section, we synthesize a sliding mode controller such that 1) the trajectory of (4.1), starting from any initial state, is globally driven onto the sliding surface (4.7) in finite time and, once the trajectory has crossed the sliding surface the first time, it will cross the surface again in every successive sampling period, resulting in a zigzag motion along the sliding surface; and 2) the size of each successive zigzagging step is non-increasing and the trajectory stays within a specified band, which is called the quasi-sliding mode band (QSMB) [212].

Now, we apply the reaching condition proposed in [212] to the following form for the SMC of system (4.1) with the sliding surface (4.7):

$$
\begin{cases}
\Delta s_i(k) = s_i(k+1) - s_i(k) \leq -\rho\lambda_i \cdot \mathrm{sgn}[s_i(k)] - \rho q_i s_i(k) \\
\qquad\qquad \text{if} \;\; s_i(k) > 0 \\
\Delta s_i(k) = s_i(k+1) - s_i(k) \geq -\rho\lambda_i \cdot \mathrm{sgn}[s_i(k)] - \rho q_i s_i(k) \\
\qquad\qquad \text{if} \;\; s_i(k) < 0
\end{cases}
\tag{4.36}
$$

where ρ represents the sampling period, $\lambda_i > 0$ and $q_i > 0$ $(i = 1, 2, \cdots, m)$ are properly chosen scalars satisfying $0 < 1 - \rho q_i < 1$, $\forall i \in \{1, 2, \cdots, m\}$. We can write (4.36) into the matrix form as follows

$$
\begin{cases}
\Delta s(k) = s(k+1) - s(k) \leq -\rho\Lambda \cdot \mathrm{sgn}[s(k)] - \rho Q s(k) \\
\qquad\qquad \text{if} \;\; s_i(k) > 0 \\
\Delta s(k) = s(k+1) - s(k) \geq -\rho\Lambda \cdot \mathrm{sgn}[s(k)] - \rho Q s(k) \\
\qquad\qquad \text{if} \;\; s_i(k) < 0
\end{cases}
\tag{4.37}
$$

with

$$
\Lambda = \mathrm{diag}\,\{\lambda_1, \lambda_2, \cdots, \lambda_m\} \in \mathbb{R}^{m \times m}, \quad Q = \mathrm{diag}\,\{q_1, q_2, \cdots, q_m\} \in \mathbb{R}^{m \times m}.
$$

Since the parameter uncertainty ΔA and the external disturbance $f(x(k))$ are both assumed to be bounded in the Euclidean norm, $\Delta_a \triangleq G\Delta A x(k)$ and $\Delta_f \triangleq GBf(x(k))$ will also be bounded. Denote δ_a^i and δ_f^i as the ith element in Δ_a and Δ_f, respectively. Suppose the lower and upper bounds on Δ_a and Δ_f are known and given as follows:

$$
\begin{aligned}
\delta_{aL}^i &\leq \delta_a^i \leq \delta_{aU}^i, \\
\delta_{fL}^i &\leq \delta_f^i \leq \delta_{fU}^i, \qquad i = 1, 2, \cdots, m
\end{aligned}
\tag{4.38}
$$

where δ_{aL}^i δ_{aU}^i δ_{fL}^i and δ_{fU}^i are all known constants. Furthermore, we denote

$$
\widehat{\Delta}_a = \begin{bmatrix} \widehat{\delta}_a^1 & \widehat{\delta}_a^2 & \cdots & \widehat{\delta}_a^m \end{bmatrix}^{\mathrm{T}},
$$

$$
\widehat{\delta}_a^i = \frac{\delta_{aU}^i + \delta_{aL}^i}{2},
$$

$$
\widetilde{\Delta}_a = \mathrm{diag}\,\{\widetilde{\delta}_a^1, \widetilde{\delta}_a^2, \cdots, \widetilde{\delta}_a^m\},
$$

$$
\widetilde{\delta}_a^i = \frac{\delta_{aU}^i - \delta_{aL}^i}{2},
$$

$$\widehat{\Delta}_f = \begin{bmatrix} \widehat{\delta}_f^1 & \widehat{\delta}_f^2 & \cdots & \widehat{\delta}_f^m \end{bmatrix}^{\mathrm{T}},$$

$$\widehat{\delta}_f^i = \frac{\delta_{fU}^i + \delta_{fL}^i}{2},$$

$$\widetilde{\Delta}_a = \text{diag}\left\{\widetilde{\delta}_f^1, \widetilde{\delta}_f^2, \cdots, \widetilde{\delta}_f^m\right\}, \tag{4.39}$$

$$\widetilde{\delta}_f^i = \frac{\delta_{fU}^i - \delta_{fL}^i}{2}, \qquad i = 1, 2, \cdots, m.$$

Now we are ready to give the design technique of the robust SMC controller.

Theorem 4.2 *Consider the uncertain nonlinear networked system (4.1) with the sliding surface (4.7) where $G = B^{\mathrm{T}}P$ and P is the solution to (4.16). If the SMC law is given as*

$$
\begin{aligned}
u(k) = &- (GB)^{-1}\Big(\rho\Lambda \cdot \text{sgn}[s(k)] + \rho Q s(k) + GAx(k-1) \\
&- Gx(k) + (\widetilde{\Delta}_a + \widehat{\Delta}_f) + (\widetilde{\Delta}_a + \widetilde{\Delta}_f)\text{sgn}[s(k)]\Big),
\end{aligned} \tag{4.40}
$$

then the state trajectories of the system (4.1) are driven onto the sliding surface with probability 1.

Proof *By (4.40), with the switching function defined in (4.7), we obtain*

$$
\begin{aligned}
\Delta s(k) =& s(k+1) - s(k) \\
=& G\Delta Ax(k) + GB\big(u(k) + f(x(k))\big) - Gx(k) + GAx(k-1) \\
=& -\rho\Lambda \cdot \text{sgn}[s(k)] - \rho Q s(k) + G\Delta Ax(k) \\
&- \Big(\widehat{\Delta}_a + \widetilde{\Delta}_a \text{sgn}[s(k)]\Big) + GBf(x(k)) \\
&- \Big(\widehat{\Delta}_f + \widetilde{\Delta}_f \text{sgn}[s(k)]\Big).
\end{aligned} \tag{4.41}
$$

Noting (4.38) and (4.39), we arrive at the reaching condition (4.37). Therefore, the state trajectories are globally driven onto the specified sliding surface. The proof is complete.

Remark 4.3 *We should point out that it is not difficult to extend the present results to systems that include polytopic parameter uncertainties, systems with stochastic disturbances, or systems with more general switching communication delays mentioned in Remark 4.2, by using the approach developed in [166, 167] and as long as the same LMI framework can be established. We could also use more up-to-date delay-dependent analysis approaches to reduce possible conservatism without difficulties. The reason why we discuss system (4.1) is to make our theory more understandable and to avoid unnecessarily complicated notations.*

4.4   An Illustrative Example

In this section, we present an illustrative example to demonstrate the effectiveness of the proposed algorithm.

Consider the following nonlinear networked system with communication delays:

$$
\begin{cases}
x(k+1) = \left(\begin{bmatrix} 0.075 & -0.5 & 0 \\ 0 & 0.08 & 0 \\ 0.01 & 0 & -0.055 \end{bmatrix} \right. \\
\qquad + \begin{bmatrix} 0.01 \\ 0.02 \\ 0 \end{bmatrix} \sin(0.6k) \begin{bmatrix} 0 & 0.01 & 0 \end{bmatrix} \left. \right) x(k) \\
\qquad + \sum_{i=1}^{3} \begin{bmatrix} 0.03 & 0 & -0.01 \\ 0.02 & 0.03 & 0 \\ 0.04 & 0.05 & -0.01 \end{bmatrix} M_i x(k - \alpha_i(k)\tau_i(k)) \\
\qquad + \begin{bmatrix} -0.085 & 0.02 \\ 0.07 & 0.05 \\ 0.02 & -0.2 \end{bmatrix} \big(u(k) + f(x(k)) \big), \\
x(k) = \varphi(k) = 0, \quad k = -d_M, -d_M + 1, \cdots, 0.
\end{cases}
\tag{4.42}
$$

Let the nonlinear function be $f(x(k)) = 0.3\sin(x(k))$. In order to design the explicit SMC controller, we suppose $G\Delta Ax(k)$ and $GBf(x(k))$ are bounded by the following conditions:

$$
\begin{aligned}
\delta_{aL}^{j} &= 0, \\
\delta_{aU}^{j} &= \|GH\|\|Ex(k)\|, \\
\delta_{fL}^{j} &= 0, \\
\delta_{fU}^{j} &= 0.3\|GB\sin(x(k))\|
\end{aligned}
\tag{4.43}
$$

where δ_{aL}^{j}, δ_{aU}^{j}, δ_{fL}^{j} and δ_{fU}^{j} ($j = 1, 2$) are defined in (4.38).

Assume that the time-varying communication delays satisfy $2 \le \tau_i(k) \le 8$ ($i = 1, 2, 3$) and

$$
\bar{\alpha}_1 = \mathbb{E}\{\alpha_1(k)\} = 0.8, \quad \bar{\alpha}_2 = \mathbb{E}\{\alpha_2(k)\} = 0.6, \quad \bar{\alpha}_3 = \mathbb{E}\{\alpha_3(k)\} = 0.4.
$$

Then, we can obtain

$$
\bar{\beta}_1 = 0.2, \quad \bar{\beta}_2 = 0.4, \quad \bar{\beta}_3 = 0.6,
$$
$$
\sigma_{\tilde{\alpha}_1}^2 = \sigma_{\tilde{\beta}_1}^2 = 0.16, \quad \sigma_{\tilde{\alpha}_2}^2 = \sigma_{\tilde{\beta}_2}^2 = 0.24, \quad \sigma_{\tilde{\alpha}_3}^2 = \sigma_{\tilde{\beta}_3}^2 = 0.24.
$$

Using the MatLab LMI toolbox to solve problem (4.35), we have

$$P = \begin{bmatrix} 5.0658 & -0.1013 & -0.1374 \\ -0.1013 & 9.4697 & -0.0181 \\ -0.1374 & -0.0181 & 8.9756 \end{bmatrix},$$

$$Q_1 = \begin{bmatrix} 0.1929 & 0.0236 & -0.0042 \\ 0.0236 & 0.0975 & -0.0003 \\ -0.0042 & -0.0003 & 0.3157 \end{bmatrix},$$

$$Q_2 = \begin{bmatrix} 0.1707 & 0.0236 & -0.0042 \\ 0.0236 & 0.1168 & -0.0003 \\ -0.0042 & -0.0003 & 0.3157 \end{bmatrix},$$

$$Q_3 = \begin{bmatrix} 0.1707 & 0.0236 & -0.0042 \\ 0.0236 & 0.0975 & -0.0003 \\ -0.0042 & -0.0003 & 0.3165 \end{bmatrix},$$

and $\gamma = 1.350392 \times 10^{-11}$ (hence the constraint $GA_d = 0$ is satisfied). Then, choosing $\rho = 0.05$, $\lambda_j = 1$ and $q_j = 1$ $(j = 1, 2)$, it follows from Theorem 4.2 that the desired SMC law can be expressed with all known parameters. The simulation results are shown in Fig. 4.1 to Fig. 4.10, which confirm that the desired requirements are well achieved.

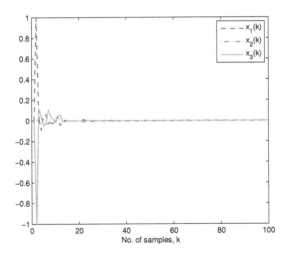

FIGURE 4.1: The trajectories of state x_k $(\rho = 0.05)$.

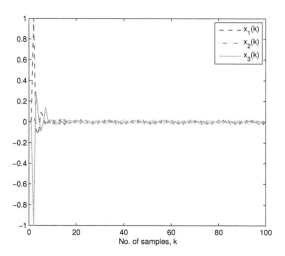

FIGURE 4.2: The trajectories of state x_k ($\rho = 0.2$).

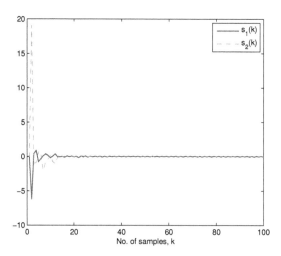

FIGURE 4.3: The trajectories of sliding variable s_k ($\rho = 0.05$).

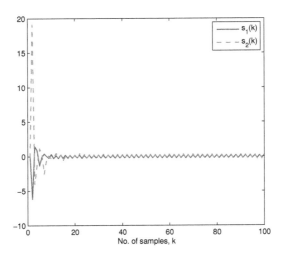

FIGURE 4.4: The trajectories of sliding variable s_k ($\rho = 0.2$).

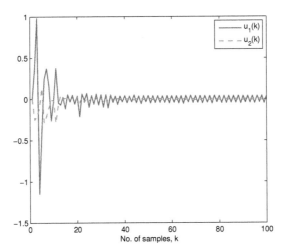

FIGURE 4.5: The control signals u_k ($\rho = 0.05$).

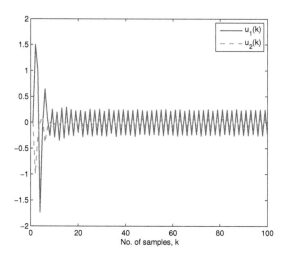

FIGURE 4.6: The control signals u_k $(\rho = 0.2)$.

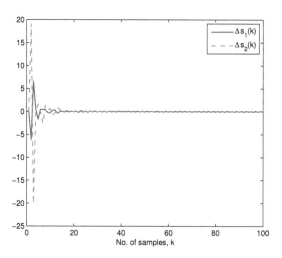

FIGURE 4.7: The signal Δs_k $(\rho = 0.05)$.

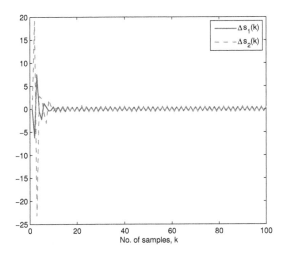

FIGURE 4.8: The signal Δs_k ($\rho = 0.2$).

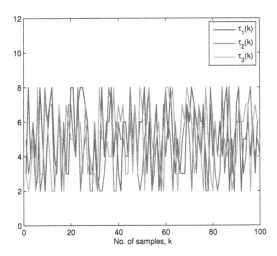

FIGURE 4.9: The time-varying communication delays $\tau(k)$.

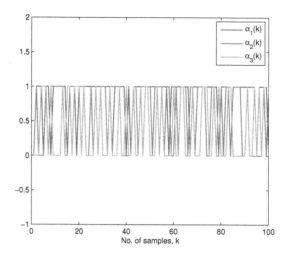

FIGURE 4.10: The Bernoulli sequences $\alpha_i(k)$ $(i = 1, 2, 3)$.

4.5   Summary

A robust SMC design problem for a type of uncertain nonlinear networked system with multiple stochastic communication delays has been studied. A new form of discrete switching function has been put forward for the first time in this chapter. By means of LMIs, sufficient conditions for the globally mean square asymptotic stability of the system on the specified sliding surface have been derived. By the reaching condition applied in this chapter, an SMC controller has been designed that globally drives the state trajectories onto the specified surface with probability 1. An illustrative numerical example has been given to show the applicability and effectiveness of the proposed method in this chapter.

5

Reliable $\mathcal{H}_\infty$ Control for A Class of Nonlinear Time-Varying Stochastic Systems with Randomly Occurring Sensor Failures

It is widely recognized that in almost all engineering applications, nonlinearities are inevitable and could not be eliminated thoroughly. Hence, nonlinear systems have gained more and more research attention, and many results have been published. On the other hand, due to the wide appearance of stochastic phenomena in almost every aspect of our daily life, stochastic systems which have found successful applications in many branches of science and engineering practice have stirred quite a lot of research interest during the past few decades. Therefore, the control as well as filtering problems for stochastic nonlinear systems have been studied extensively so as to meet the ever-increasing demand for systems with both stochasticity and nonlinearities.

On the other hand, in real-world engineering practice, it is well known that sensors have always been confronted with different kinds of failures due to a variety of reasons, including the erosion caused by severe circumstance, abrupt changes of working conditions, intense external disturbance and internal component constraints and aging. As requirements increase for the reliability of engineering systems, the fault-tolerant control problem, also known as the reliable control problem which aims to stabilize the systems accurately and precisely in spite of possible failures, has therefore attracted considerable attention in the past decades. It is worth pointing out that, in many real-world engineering practices, due to random abrupt changes of environmental circumstances or working conditions, most of the time the sensors might encounter failures in a probabilistic way, that is, randomly occurring sensor failures (ROSFs). Different from the traditional deterministic sensor failure model, such randomly occurring sensor failure could be either existent or non-existent at a specific time and the existence is governed by certain Bernoulli distributed white sequences with known probabilities. However, in spite of its clear physical insight and importance in engineering application, the control problem for nonlinear stochastic systems under the circumstance of randomly occurring sensor failures has not yet been studied sufficiently.

Motivated by the above discussion, in this chapter, we aim to design an output feedback controller for a class of time-varying nonlinear stochastic systems with an $\mathcal{H}_\infty$ specification in the presence of randomly occurring sensor failures.

The rest of the chapter is arranged as follows. Section 5.1 formulates the $\mathcal{H}_\infty$ control problem for the time-varying nonlinear stochastic systems with ROSF. In Section 5.2, $\mathcal{H}_\infty$ performance is analyzed in terms of certain LMIs. Section 5.3 gives the methodology to solve the addressed control problem and outlines the computational algorithm to recursively obtain the required controller parameters. A numerical example is presented in Section 5.4 to show the effectiveness and applicability of the proposed algorithm. Section 5.5 draws the conclusion.

5.1   Problem Formulation

Consider the following time-varying nonlinear stochastic system defined on $k \in [0, N-1]$:

$$\begin{cases} x(k+1) = A(k)x(k) + B(k)u(k) + f\big(x(k), k\big) + D(k)\omega(k) \\ y(k) = \varphi\big(C(k)x(k)\big) + g\big(x(k), k\big) + E(k)\omega(k) \end{cases} \quad (5.1)$$

where $x(k)$, $y(k)$ and $u(k)$ represent the system state, output and control input, respectively. $\omega(k)$ represents the external disturbance belonging to l_2. $A(k)$, $B(k)$, $C(k)$, $D(k)$ and $E(k)$ are known real time-varying matrices with appropriate dimensions.

The nonlinear stochastic functions $f\big(x(k), k\big)$ and $g\big(x(k), k\big)$ are assumed to have the following first moments for all $x(k)$ and k:

$$\mathbb{E}\left\{ \begin{bmatrix} f\big(x(k), k\big) \\ g\big(x(k), k\big) \end{bmatrix} \middle| x(k) \right\} = 0, \quad (5.2)$$

with the covariance given by

$$\mathbb{E}\left\{ \begin{bmatrix} f\big(x(k), k\big) \\ g\big(x(k), k\big) \end{bmatrix} \begin{bmatrix} f^{\mathrm{T}}\big(x(j), j\big) & g^{\mathrm{T}}\big(x(j), j\big) \end{bmatrix} \middle| x(k) \right\} = 0, \quad k \neq j \quad (5.3)$$

and

$$\mathbb{E}\left\{ \begin{bmatrix} f\big(x(k), k\big) \\ g\big(x(k), k\big) \end{bmatrix} \begin{bmatrix} f^{\mathrm{T}}\big(x(k), k\big) & g^{\mathrm{T}}\big(x(k), k\big) \end{bmatrix} \middle| x(k) \right\}$$

$$= \sum_{i=1}^{q} \Omega^i(k)\Big(x^{\mathrm{T}}(k)\Gamma_i(k)x(k) \Big) \quad (5.4)$$

$$= \sum_{i=1}^{q} \begin{bmatrix} \Omega^i_{11}(k) & \Omega^i_{12}(k) \\ (\Omega^i_{12})^{\mathrm{T}}(k) & \Omega^i_{22}(k) \end{bmatrix} \Big(x^{\mathrm{T}}(k)\Gamma_i(k)x(k) \Big),$$

where $\Omega_{11}^i(k) > 0$, $\Omega_{22}^i(k) > 0$, $\Omega_{12}^i(k)$ and $\Gamma_i(k) > 0$ $(i = 1, 2, \cdots, q)$ are known matrices with compatible dimensions.

Remark 5.1 *As discussed in [240], the stochastic nonlinearities* $f(x(k), k)$ *and* $g(x(k), k)$ *account for several classes of well-studied nonlinear systems, such as the system with state-dependent multiplicative noises and the system whose states have power dependent on the sign of the nonlinear function of the state.*

The nonlinear function $\varphi(\cdot) : \mathbb{R}^m \mapsto \mathbb{R}^m$ which illustrates the phenomenon of randomly occurring sensor failures is defined as follows:

$$\varphi\big(C(k)x(k)\big) = \alpha(k)\kappa\big(C(k)x(k)\big) + \big(1 - \alpha(k)\big)C(k)x(k) \qquad (5.5)$$

where the sensor failures function $\kappa(\cdot) : \mathbb{R}^m \mapsto \mathbb{R}^m$ has the following form:

$$\kappa\big(C(k)x(k)\big) = F\big(x(k), k\big)C(k)x(k) \qquad (5.6)$$

with

$$F(x(k), k) = \mathrm{diag}\{f_1(k), f_2(k), \ldots, f_m(k)\},$$
$$0 \le f_{il}(k) \le f_i(k) \le f_{iu}(k) < \infty, f_{il}(k) < 1, f_{iu}(k) \ge 1. \qquad (5.7)$$

Remark 5.2 *The time-varying equation (5.6) is used to interpret the constraints on the output sampling when the sensor failures phenomenon appears. Such a model could explain several types of sensor failures which frequently appear in the engineering practice including output signal quantization in [193] and sensor saturation in [235].*

The random variable $\alpha(k)$ in (5.5) which accounts for the randomly occurring sensor failures is a Bernoulli distributed white sequence taking values on 0 and 1 with

$$\mathcal{P}\{\alpha(k) = 1\} = \mathbb{E}\{\alpha(k)\} = \bar{\alpha} \qquad (5.8)$$

with $\bar{\alpha}$ being a known positive constant. Hence, it can be easily obtained that

$$\sigma_\alpha^2 = \mathbb{E}\{(\alpha(k) - \bar{\alpha})^2\} = (1 - \bar{\alpha})\bar{\alpha}. \qquad (5.9)$$

Moreover, $\alpha(k)$ is assumed to be independent of $\omega(k)$ and the stochastic nonlinearities $f(x(k), k)$ and $g(x(k), k)$.

Remark 5.3 *The stochastic variable* $\alpha(k)$ *is introduced to illustrate the phenomenon of randomly occurring sensor failures. In such a model, when* $\alpha(k) = 0$, *it means the sensor is in good condition; otherwise there might be sensor failures which will bring about certain constraints on the output sampling. Compared to the traditional models where the sensor failures are assumed to happen in a deterministic way, the model we proposed here has the advantage of describing the phenomenon that due to time-varying working conditions like the change of circumstance, or the instability of the signal transmission channels, the sensor fault might be varying during a known band obeying certain probabilistic distribution. Therefore, such a description is more realistic in engineering practice and less conservative than those in existing literature.*

Let $F\big(x(k),k\big) = H(k) + H(k)N(k)$, where

$$H(k) = \text{diag}\{h_1(k), h_2(k), \ldots, h_m(k)\},$$
$$h_i(k) = \frac{f_{il}(k) + f_{iu}(k)}{2},$$
$$N(k) = \text{diag}\{n_1(k), n_2(k), \ldots, n_m(k)\}, \tag{5.10}$$
$$n_i(k) = \frac{f_i(k) - h_i(k)}{h_i(k)}.$$

Denoting $L(k) = \text{diag}\{l_1(k), l_2(k), \ldots, l_m(k)\}$ with $l_i(k) = (f_{iu}(k) - f_{il}(k))/(f_{iu}(k) + f_{il}(k))$, we obtain

$$N^{\mathrm{T}}(k)N(k) \le L^{\mathrm{T}}(k)L(k) \le I. \tag{5.11}$$

So far, we have converted the influence of the sensor failures on the system output sampling into certain norm-bounded parameter uncertainties. Then the original system (5.1) can be re-formulated as follows:

$$\begin{cases} x(k+1) = A(k)x(k) + B(k)u(k) + f\big(x(k),k\big) + D(k)w(k) \\ y(k) = \hat{C}(k)x(k) + \alpha(k)H(k)N(k)C(k)x(k) \\ \qquad + g\big(x(k),k\big) + E(k)w(k) \end{cases} \tag{5.12}$$

where $\hat{C}(k) = \big(\alpha(k)H(k) + (1 - \alpha(k))I\big)C(k)$.

In the following, we give the design objective of this chapter. First, we define the $\mathcal{H}_\infty$ performance index for system (5.12) as follows:

$$J_\infty = \mathbb{E}\left\{ \sum_{k=0}^{N-1} \big(\|y(k)\|^2 - \gamma^2\|w(k)\|^2\big) \right\} - \gamma^2\|x(0)\|_W^2 \tag{5.13}$$

where W is a known weighted matrix.

In this chapter, it is our objective to determine an output feedback controller of the following form:

$$\begin{cases} x_g(k+1) = A_g(k)x_g(k) + B_g(k)y(k), \\ \qquad u(k) = C_g(k)x_g(k), \end{cases} \tag{5.14}$$

such that, for the closed-loop system, given a pre-specified $\mathcal{H}_\infty$ disturbance attenuation level $\gamma > 0$, $J_\infty < 0$ holds for all non-zero $w(k)$ and $x(0)$.

Before giving our main results, first, we need to reconstruct the system (5.12) so as to obtain a system without parameter uncertainty $N(k)$ as follows:

$$\begin{cases} x(k+1) = A(k)x(k) + B(k)u(k) + f\big(x(k),k\big) + \hat{D}(k)\hat{w}(k) \\ y(k) = \hat{C}(k)x(k) + g\big(x(k),k\big) + \hat{E}(k)\hat{w}(k) \end{cases} \tag{5.15}$$

where

$$\hat{\omega}(k) = \left[\begin{array}{c} \omega(k) \\ \gamma^{-1}\epsilon(k)N(k)C(k)x(k) \end{array} \right],$$

$$\hat{D}(k) = \left[\begin{array}{cc} D(k) & 0 \end{array} \right],$$

$$\hat{E}(k) = \left[\begin{array}{cc} E(k) & \gamma\alpha(k)\epsilon^{-1}(k)H(k) \end{array} \right]$$

with $\epsilon(k) \neq 0$ being any known real-valued constant.

We now consider the following performance index:

$$\bar{J}_\infty = \mathbb{E} \left\{ \sum_{k=0}^{N-1} \left(\|y(k)\|^2 + \|\epsilon(k)L(k)C(k)x(k)\|^2 - \gamma^2 \|\hat{\omega}(k)\|^2 \right) \right\} \tag{5.16}$$
$$- \gamma^2 \|x(0)\|_W^2.$$

Next, we shall show that $\bar{J}_\infty$ is an upper bound of J_∞. To this end, subtracting (5.13) from (5.16) leads to

$$\bar{J}_\infty - J_\infty$$

$$= \mathbb{E} \left\{ \sum_{k=0}^{N-1} \|\epsilon(k)L(k)C(k)x(k)\|^2 - \gamma^2 \left(\|\hat{\omega}(k)\|^2 - \|\omega(k)\|^2 \right) \right\}$$

$$= \mathbb{E} \left\{ \sum_{k=0}^{N-1} \|\epsilon(k)L(k)C(k)x(k)\|^2 - \|\epsilon(k)N(k)C(k)x(k)\|^2 \right\} \tag{5.17}$$

$$= \mathbb{E} \left\{ \sum_{k=0}^{N-1} \|\epsilon(k)\sqrt{L^{\mathrm{T}}(k)L(k) - N^{\mathrm{T}}(k)N(k)}C(k)x(k)\|^2 \right\}$$

$$\geq 0.$$

Hence, $\bar{J}_\infty$ is an upper bound of J_∞. It is worth noting that since $N(k)$ can be any arbitrary matrix corresponding to any arbitrary sensor failures, $\bar{J}_\infty$ is a tight upper bound of J_∞.

Applying the full order dynamic output controller (5.14) to system (5.15), we have the following closed-loop system:

$$\left\{ \begin{array}{l} \xi(k+1) = A_f(k)\xi(k) + G_f h(k) + D_f \hat{\omega}(k) \\ y(k) = C_f(k)\xi(k) + g(x(k), k) + \hat{E}(k)\hat{\omega}(k) \end{array} \right. \tag{5.18}$$

where

$$\xi(k) = \left[\begin{array}{c} x(k) \\ x_g(k) \end{array} \right],$$

$$h(k) = \left[\begin{array}{c} f(x(k), k) \\ g(x(k), k) \end{array} \right],$$

$$A_f(k) = \left[\begin{array}{cc} A(k) & B(k)C_g(k) \\ B_g(k)\hat{C}(k) & A_g(k) \end{array} \right],$$

$$G_f(k) = \left[\begin{array}{cc} I & 0 \\ 0 & B_g(k) \end{array} \right],$$

$$D_f(k) = \begin{bmatrix} \hat{D}(k) \\ B_g(k)\hat{E}(k) \end{bmatrix},$$

$$C_f(k) = \begin{bmatrix} \hat{C}(k) & 0 \end{bmatrix}.$$

Now we are in the position to deal with the reliable $\mathcal{H}_\infty$ control problem for the nonlinear stochastic system with ROSF.

5.2 $\mathcal{H}_\infty$ Performance

In this section, we first analyze the disturbance attenuation level for the time-varying nonlinear system with randomly occurring sensor failures. A theorem which plays a vital role in the controller design stage will then be presented to give a sufficient condition guaranteeing the pre-specified $\mathcal{H}_\infty$ performance index. To this end, first, for the convenience of notation, we denote

$$\bar{C}_f(k) = \begin{bmatrix} \bar{C}(k) & 0 \end{bmatrix}, \qquad \bar{C}(k) = \Big(\bar{\alpha}H(k) + (1 - \bar{\alpha})I\Big)C(k),$$

$$\check{C}_f(k) = \begin{bmatrix} \check{C}(k) & 0 \end{bmatrix}, \qquad \check{C}(k) = \sigma_\alpha\big(H(k) - I\big)C(k),$$

$$\bar{A}_f(k) = \begin{bmatrix} A(k) & B(k)C_g(k) \\ B_g(k)\bar{C}(k) & A_g(k) \end{bmatrix},$$

$$\check{A}_f(k) = \begin{bmatrix} 0 & 0 \\ B_g(k)\check{C}(k) & 0 \end{bmatrix}, \tag{5.19}$$

$$\bar{E}(k) = \begin{bmatrix} E(k) & \gamma\bar{\alpha}\epsilon^{-1}(k)H(k) \end{bmatrix},$$

$$\check{E}(k) = \begin{bmatrix} 0 & \gamma\sigma_\alpha\epsilon^{-1}(k)H(k) \end{bmatrix},$$

$$\bar{D}_f(k) = \begin{bmatrix} \hat{D}(k) \\ B_g(k)\bar{E}(k) \end{bmatrix}, \qquad \check{D}_f(k) = \begin{bmatrix} 0 \\ B_g(k)\check{E}(k) \end{bmatrix},$$

$$\bar{W} = \mathrm{diag}\{W, 0\}, \qquad Z = \begin{bmatrix} I & 0 \end{bmatrix}.$$

The following theorem gives a sufficient condition for system (5.18) capable of satisfying the pre-specified $\mathcal{H}_\infty$ requirement.

Theorem 5.1 *Given a positive scalar $\gamma > 0$, let the feedback controller gains $A_g(k)$, $B_g(k)$ and $C_g(k)$ be given. If there exists a sequence of positive definite matrices $\{P(k)\}_{1 \le k \le N}$ with $P(0) \le \gamma^2\bar{W}$ satisfying the following iterative matrix inequalities:*

$$\Lambda(k) \triangleq \begin{bmatrix} \Lambda_{11}(k) & \Lambda_{12}(k) \\ \Lambda_{12}^{\mathrm{T}}(k) & \Lambda_{22}(k) \end{bmatrix} < 0 \tag{5.20}$$

where

$$\Lambda_{11}(k) = \bar{A}_f^{\mathrm{T}}(k)P(k+1)\bar{A}_f(k) + \check{A}_f^{\mathrm{T}}(k)P(k+1)\check{A}_f(k) - P(k)$$
$$+ \bar{C}_f^{\mathrm{T}}(k)\bar{C}_f(k) + \check{C}_f^{\mathrm{T}}(k)\check{C}_f(k)$$
$$+ \sum_{i=1}^{q} Z^{\mathrm{T}}\Gamma_i(k)Z\big(\mathrm{tr}[G_f(k)^{\mathrm{T}}P(k+1)G_f(k)\Omega^i(k)] + \mathrm{tr}[\Omega_{22}^i(k)]\big)$$
$$+ \epsilon^2(k)Z^{\mathrm{T}}C^{\mathrm{T}}(k)L^{\mathrm{T}}(k)L(k)C(k)Z,$$

$$\Lambda_{12}(k) = \bar{A}_f^{\mathrm{T}}(k)P(k+1)\bar{D}_f(k) + \check{A}_f^{\mathrm{T}}(k)P(k+1)\check{D}_f(k)$$
$$+ \bar{C}_f^{\mathrm{T}}(k)\bar{E}(k) + \check{C}_f^{\mathrm{T}}(k)\check{E}(k),$$
$$\Lambda_{22}(k) = -\gamma^2 I + \bar{D}_f^{\mathrm{T}}(k)P(k+1)\bar{D}_f(k) + \check{D}_f^{\mathrm{T}}(k)P(k+1)\check{D}_f(k)$$
$$+ \bar{E}^{\mathrm{T}}(k)\bar{E}(k) + \check{E}^{\mathrm{T}}(k)\check{E}(k),$$

then the required $\mathcal{H}_\infty$ performance of the closed-loop system (5.18) is achieved, that is, $J_\infty < 0$.

Proof *By defining*

$$J_k = \xi^{\mathrm{T}}(k+1)P(k+1)\xi(k+1) - \xi^{\mathrm{T}}(k)P(k)\xi(k), \qquad (5.21)$$

we have

$$
\begin{aligned}
\mathbb{E}\{J_k\} =& \mathbb{E}\Big\{ \Big(A_f(k)\xi(k) + G_f h(k) + D_f \hat{\omega}(k)\Big)^{\mathrm{T}} P(k+1)\Big(A_f(k)\xi(k) \\
& + G_f h(k) + D_f \hat{\omega}(k)\Big)\Big\} - \xi^{\mathrm{T}}(k)P(k)\xi(k) \\
=& \mathbb{E}\{\xi^{\mathrm{T}}(k)\Big(A_f^{\mathrm{T}}(k)P(k+1)A_f(k) - P(k) \\
& + \sum_{i=1}^{q} Z^{\mathrm{T}}\Gamma_i(k)Z\mathrm{tr}[G_f(k)^{\mathrm{T}}P(k+1)G_f(k)\Omega^i(k)]\Big)\xi(k) \\
& + 2\xi^{\mathrm{T}}(k)A_f^{\mathrm{T}}(k)P(k+1)D_f(k)\hat{\omega}(k) \\
& + \hat{\omega}^{\mathrm{T}}(k)D_f^{\mathrm{T}}(k)P(k+1)D_f(k)\hat{\omega}(k)\}.
\end{aligned}
\qquad (5.22)
$$

Adding the following zero term

$$
\begin{aligned}
\|y(k)\|^2 + & \|\epsilon(k)L(k)C(k)x(k)\|^2 - \gamma^2\|\hat{\omega}(k)\|^2 - (\|y(k)\|^2 \\
& + \|\epsilon(k)L(k)C(k)x(k)\|^2 - \gamma^2\|\hat{\omega}(k)\|^2)
\end{aligned}
\qquad (5.23)
$$

to both sides of (5.22) and then taking mathematical expectation results in

$$
\begin{aligned}
\mathbb{E}\{J_k\} =& \mathbb{E}\Big\{ \xi^{\mathrm{T}}(k)\Big(A_f^{\mathrm{T}}(k)P(k+1)A_f(k) - P(k) + C_f^{\mathrm{T}}(k)C_f(k) \\
& + \epsilon^2(k)Z^{\mathrm{T}}C^{\mathrm{T}}(k)L^{\mathrm{T}}(k)L(k)C(k)Z + \sum_{i=1}^{q} Z^{\mathrm{T}}\Gamma_i(k) \\
& \times Z\big(\mathrm{tr}[G_f^{\mathrm{T}}(k)P(k+1)G_f(k)\Omega^i(k)] + \mathrm{tr}[\Omega_{22}^i(k)]\big)\Big)\xi(k) \\
& + 2\xi^{\mathrm{T}}(k)\Big(A_f^{\mathrm{T}}(k)P(k+1)D_f(k) + C_f^{\mathrm{T}}(k)\hat{E}(k)\Big)\hat{\omega}(k) \\
& + \hat{\omega}^{\mathrm{T}}(k)\Big(D_f^{\mathrm{T}}(k)P(k+1)D_f(k) + \hat{E}^{\mathrm{T}}(k)\hat{E}(k)\Big)\hat{\omega}(k) \\
& - \Big(\|y(k)\|^2 + \|\epsilon(k)L(k)C(k)x(k)\|^2 - \gamma^2\|\hat{\omega}(k)\|^2\Big)\Big\}.
\end{aligned}
\qquad (5.24)
$$

Note that there are stochastic parameters $C_f(k)$ and $\hat{E}(k)$ in (5.24). Thus, in the following, we aim to eliminate the random variables so as to obtain a deterministic condition. To this end, first, denote

$$
\begin{aligned}
\tilde{\alpha}(k) &= \alpha(k) - \bar{\alpha}, \\
C_f(k) &= \bar{C}_f(k) + \tilde{C}_f(k), \\
\tilde{C}_f(k) &= \left[\ \left(\tilde{\alpha}(k)H(k) - \tilde{\alpha}(k)I\right)C(k)\quad 0\ \right], \\
\hat{E}(k) &= \bar{E}(k) + \tilde{E}(k), \quad \tilde{E}(k) = \left[\ 0\quad \gamma\tilde{\alpha}(k)\epsilon^{-1}(k)H(k)\ \right],
\end{aligned}
\tag{5.25}
$$

where $\bar{C}_f(k)$ and $\bar{E}(k)$ are defined in (5.19).

Taking the statistical property of $\alpha(k)$ into consideration, we could obtain

$$
\begin{aligned}
\mathbb{E}\{C_f^{\mathrm{T}}(k)C_f(k)\} &= \bar{C}_f^{\mathrm{T}}(k)\bar{C}_f(k) + \mathbb{E}\{\tilde{C}_f^{\mathrm{T}}(k)\tilde{C}_f(k)\} \\
&= \bar{C}_f^{\mathrm{T}}(k)\bar{C}_f(k) + \check{C}_f^{\mathrm{T}}(k)\check{C}_f(k), \\
\mathbb{E}\{C_f^{\mathrm{T}}(k)\hat{E}(k)\} &= \bar{C}_f^{\mathrm{T}}(k)\bar{E}(k) + \mathbb{E}\{\tilde{C}_f^{\mathrm{T}}(k)\tilde{E}(k)\} \\
&= \bar{C}_f^{\mathrm{T}}(k)\bar{E}(k) + \check{C}_f^{\mathrm{T}}(k)\check{E}(k), \\
\mathbb{E}\{\hat{E}^{\mathrm{T}}(k)\hat{E}(k)\} &= \bar{E}^{\mathrm{T}}(k)\bar{E}(k) + \mathbb{E}\{\tilde{E}^{\mathrm{T}}(k)\tilde{E}(k)\} \\
&= \bar{E}^{\mathrm{T}}(k)\bar{E}(k) + \check{E}^{\mathrm{T}}(k)\check{E}(k),
\end{aligned}
\tag{5.26}
$$

where $\check{C}_f(k)$ and $\check{E}(k)$ are defined in (5.19).

Therefore, it can be seen that

$$
\begin{aligned}
\mathbb{E}\{J_k\} =&\, \xi^{\mathrm{T}}(k)\Big(\bar{A}_f^{\mathrm{T}}(k)P(k+1)\bar{A}_f(k) + \check{A}_f^{\mathrm{T}}(k)P(k+1)\check{A}_f(k) \\
&- P(k) + \bar{C}_f^{\mathrm{T}}(k)\bar{C}_f(k) + \check{C}_f^{\mathrm{T}}(k)\check{C}_f(k) \\
&+ \epsilon^2(k)Z^{\mathrm{T}}C^{\mathrm{T}}(k)L^{\mathrm{T}}(k)L(k)C(k)Z + \sum_{i=1}^{q} Z^{\mathrm{T}}\Gamma_i(k) \\
&\times Z\big(\mathrm{tr}[G_f^{\mathrm{T}}(k)P(k+1)G_f(k)\Omega^i(k)] + \mathrm{tr}[\Omega_{22}^i(k)]\big)\Big)\xi(k) \\
&+ 2\xi^{\mathrm{T}}(k)\Big(\bar{A}_f^{\mathrm{T}}(k)P(k+1)\bar{D}_f(k) + \check{A}_f^{\mathrm{T}}(k)P(k+1)\check{D}_f(k) \\
&+ \bar{C}_f^{\mathrm{T}}(k)\bar{E}(k) + \check{C}_f^{\mathrm{T}}(k)\check{E}(k)\Big)\hat{\omega}(k) + \hat{\omega}^{\mathrm{T}}(k)\Big(-\gamma^2 I \\
&+ \bar{D}_f^{\mathrm{T}}(k)P(k+1)\bar{D}_f(k) + \check{D}_f^{\mathrm{T}}(k)P(k+1)\check{D}_f(k) \\
&+ \bar{E}^{\mathrm{T}}(k)\bar{E}(k) + \check{E}^{\mathrm{T}}(k)\check{E}(k)\Big)\hat{\omega}(k) \\
&- \mathbb{E}\{\|y(k)\|^2 + \|\epsilon(k)L(k)C(k)x(k)\|^2 - \gamma^2\|\hat{\omega}(k)\|^2\} \\
=&\, \left[\begin{array}{c}\xi(k) \\ \hat{\omega}(k)\end{array}\right]^{\mathrm{T}} \Lambda(k) \left[\begin{array}{c}\xi(k) \\ \hat{\omega}(k)\end{array}\right] - \mathbb{E}\{\|y(k)\|^2 \\
&+ \|\epsilon(k)L(k)C(k)x(k)\|^2 - \gamma^2\|\hat{\omega}(k)\|^2\}.
\end{aligned}
\tag{5.27}
$$

It then follows that

$$
\mathbb{E}\{\sum_{k=0}^{N-1} J_k\} = \mathbb{E}\{\xi^{\mathrm{T}}(N)P(N)\xi(N)\} - \xi^{\mathrm{T}}(0)P(0)\xi(0)
$$

$$
= \sum_{k=0}^{N-1} \left(\left[\begin{array}{c} \xi(k) \\ \hat{\omega}(k) \end{array} \right]^{\mathrm{T}} \Lambda(k) \left[\begin{array}{c} \xi(k) \\ \hat{\omega}(k) \end{array} \right] \right. \tag{5.28}
$$

$$
\left. + \mathbb{E}\{\|y(k)\|^2 + \|\epsilon(k)L(k)C(k)x(k)\|^2 - \gamma^2\|\hat{\omega}(k)\|^2\} \right),
$$

which indicates

$$
\bar{J}_\infty = \sum_{k=0}^{N-1} \left(\mathbb{E}\{\|y(k)\|^2 + \|\epsilon(k)L(k)C(k)x(k)\|^2 - \gamma^2\|\hat{\omega}(k)\|^2\} \right)
$$

$$
\quad - \gamma^2\|x(0)\|_W^2
$$

$$
= \xi^{\mathrm{T}}(0)\big(P(0) - \gamma^2\bar{W}\big)\xi(0) - \mathbb{E}\{\xi^{\mathrm{T}}(N)P(N)\xi(N)\} \tag{5.29}
$$

$$
\quad + \sum_{k=0}^{N-1} \left(\left[\begin{array}{c} \xi(k) \\ \hat{\omega}(k) \end{array} \right]^{\mathrm{T}} \Lambda(k) \left[\begin{array}{c} \xi(k) \\ \hat{\omega}(k) \end{array} \right] \right).
$$

Therefore, noticing $P(0) \leq \gamma^2\bar{W}$, $P(N) > 0$ and $\Lambda(k) < 0$, we arrive at

$$
\bar{J}_\infty < 0, \tag{5.30}
$$

which means $J_\infty < 0$. The proof is complete.

In this section, we have analyzed the $\mathcal{H}_\infty$ performance of the closed-loop system, and obtained an important sufficient condition which will play a key role in the controller design procedure which is shown in the following section.

5.3   Controller Design

In this section, we aim to find the desired controller that is able to satisfy the required $\mathcal{H}_\infty$ performance. We shall present a theorem to give the sufficient condition of the existence of such a controller. Then a computational algorithm will be proposed to obtain the numerical values of the feedback gains at each sampling instant k.

Controller Design

Before giving the controller design algorithm, we first introduce the following lemma which is useful in the derivation later.

Lemma 5.1 *(Schur Complement Lemma) Given constant matrices S_1, S_2, S_3 where $S_1 = S_1^T$ and $0 < S_2 = S_2^T$, then $S_1 + S_3^T S_2^{-1} S_3 < 0$ if and only if*

$$
\begin{bmatrix} S_1 & S_3^T \\ S_3 & -S_2 \end{bmatrix} < 0 \quad \text{or} \quad \begin{bmatrix} -S_2 & S_3 \\ S_3^T & S_1 \end{bmatrix} < 0. \tag{5.31}
$$

The following theorem provides a sufficient condition for the existence of the desired controller of the time-varying nonlinear stochastic system (5.1) with randomly occurring sensor failures.

Theorem 5.2 *For the pre-specified disturbance attenuation level $\gamma > 0$, if, over a finite horizon $[0, N)$, there exist sequences of positive definite matrices $Q_1(k)$, $Q_2(k)$, real-valued matrices $Q_3(k)$, positive scalars $\beta_i(k)$ ($i = 1, 2, \ldots, q$) and sequences of matrices $A_g(k)$, $B_g(k)$ and $C_g(k)$ such that the following LMIs:*

$$
\begin{bmatrix} -\beta_i(k) & \theta_{1i}^T(k) & \theta_{2i}^T(k)B_g^T(k) \\ * & -Q_1(k+1) & -Q_3(k+1) \\ * & * & -Q_2(k+1) \end{bmatrix} < 0, \tag{5.32}
$$

$$
\begin{bmatrix}
\tilde{\Lambda}_{11}(k) & -P_3(k) & 0 & \bar{C}^T(k) & \check{C}^T(k) \\
* & -P_2(k) & 0 & 0 & 0 \\
* & * & -\gamma^2 I & \bar{E}^T(k) & \check{E}^T(k) \\
* & * & * & -I & 0 \\
* & * & * & * & -I \\
* & * & * & * & * \\
* & * & * & * & * \\
* & * & * & * & *
\end{bmatrix}
$$

$$
\begin{matrix}
A^T(k) & \bar{C}^T(k)B_g^T(k) & 0 & \check{C}^T(k)B_g^T(k) \\
C_g^T(k)B^T(k) & A_g^T(k) & 0 & 0 \\
\hat{D}^T(k) & \bar{E}^T(k)B_g^T(k) & 0 & \check{E}^T(k)B_g^T(k) \\
0 & 0 & 0 & 0 \\
0 & 0 & 0 & 0 \\
-Q_1(k+1) & -Q_3(k+1) & 0 & 0 \\
* & -Q_2(k+1) & 0 & 0 \\
* & * & -Q_1(k+1) & -Q_3(k+1) \\
* & * & * & -Q_2(k+1)
\end{matrix} < 0 \tag{5.33}
$$

where

$$
\tilde{\Lambda}_{11}(k) = -P_1(k) + \sum_{i=1}^{q} \Gamma_i(k)\big(\beta_i(k) + \text{tr}[\Omega_{22}^i(k)]\big) + \epsilon^2(k)C^T(k)L^T(k)L(k)C(k)
$$

are satisfied with the initial condition

$$
\begin{cases} P_1(0) \leq \gamma^2 W \\ P_2(0) = P_3(0) = 0 \end{cases} \tag{5.34}
$$

and the following parameters update formula

$$
\begin{cases}
P_1(k+1) = \left(Q_1(k+1) - Q_3(k+1)Q_2^{-1}(k+1)Q_3^{\mathrm{T}}(k+1)\right)^{-1}, \\
P_2(k+1) = \left(Q_2(k+1) - Q_3^{\mathrm{T}}(k+1)Q_1^{-1}(k+1)Q_3(k+1)\right)^{-1}, \\
P_3(k+1) = -\, Q_1^{-1}(k+1)Q_3(k+1)\left(Q_2(k+1)\right. \\
\qquad\qquad \left. -\, Q_3^{\mathrm{T}}(k+1)Q_1^{-1}(k+1)Q_3(k+1)\right)^{-1},
\end{cases}
\tag{5.35}
$$

then the addressed reliable $\mathcal{H}_\infty$ controller exists for the nonlinear stochastic system (5.1). The controller gains $A_g(k)$, $B_g(k)$ and $C_g(k)$ can be obtained by solving the corresponding set of LMIs at time point k.

Proof *Suppose the variable $P(k)$ can be decomposed as follows:*

$$
P(k) = \begin{bmatrix} P_1(k) & P_3(k) \\ P_3^{\mathrm{T}}(k) & P_2(k) \end{bmatrix}, \qquad
P^{-1}(k) = \begin{bmatrix} Q_1(k) & Q_3(k) \\ Q_3^{\mathrm{T}}(k) & Q_2(k) \end{bmatrix}.
\tag{5.36}
$$

Moreover, not losing generality, it is assumed that $\Omega^i(k)$ has the following form:

$$
\Omega^i(k) \triangleq \theta_i(k)\theta_i^{\mathrm{T}}(k) = \begin{bmatrix} \theta_{1i}(k) \\ \theta_{2i}(k) \end{bmatrix} \begin{bmatrix} \theta_{1i}(k) \\ \theta_{2i}(k) \end{bmatrix}^{\mathrm{T}}
\tag{5.37}
$$

with $\theta_{1i}(k)$ and $\theta_{2i}(k)$ ($i = 1, 2, \ldots, q$) being vectors of appropriate dimensions. Using Lemma 5.1, we could see that the following inequality

$$
\mathrm{tr}[G_f^{\mathrm{T}}(k)P(k+1)G_f(k)\Omega^i(k)] < \beta_i(k)
\tag{5.38}
$$

is equivalent to

$$
\begin{bmatrix} -\beta_i(k) & \theta_i^{\mathrm{T}}(k)G_f^{\mathrm{T}}(k) \\ G_f(k)\theta_i(k) & -P^{-1}(k+1) \end{bmatrix} < 0.
\tag{5.39}
$$

Hence, after some calculation, we find out that inequality (5.38) is equivalent to LMI (5.32).

Now, let us pay attention to LMI (5.33). Again, by Schur Complement Equivalence, it can be seen that the LMI (5.33) is true if and only if the following inequality is true:

$$
\begin{bmatrix}
\bar{\Lambda}_{11}(k) & 0 & \bar{A}_f^{\mathrm{T}}(k) & \check{A}_f^{\mathrm{T}}(k) \\
* & -\gamma^2 I + \bar{E}^{\mathrm{T}}(k)\bar{E}(k) + \check{E}^{\mathrm{T}}(k)\check{E}(k) & \bar{D}_f^{\mathrm{T}}(k) & \check{D}_f^{\mathrm{T}}(k) \\
* & * & -P^{-1}(k+1) & 0 \\
* & * & * & -P^{-1}(k+1)
\end{bmatrix} < 0,
\tag{5.40}
$$

where

$$\bar{\Lambda}_{11}(k) = -P(k) + \bar{C}_f^{\mathrm{T}}(k)\bar{C}_f(k) + \check{C}_f^{\mathrm{T}}(k)\check{C}_f(k)$$

$$+ \sum_{i=1}^{q} Z^{\mathrm{T}}\Gamma_i(k)Z\big(\beta_i(k) + \mathrm{tr}[\Omega_{22}^i(k)]\big)$$

$$+ \epsilon^2(k)Z^{\mathrm{T}}C^{\mathrm{T}}(k)L^{\mathrm{T}}(k)L(k)C(k)Z.$$

Next, note that due to inequality (5.38), the matrix inequality (5.20) in Theorem 1 is implied by LMI (5.33), which means the $\mathcal{H}_\infty$ performance of the closed-loop system is achieved. The proof is complete.

Remark 5.4 *The proposed theorem gives a sufficient condition for the existence of the required reliable $\mathcal{H}_\infty$ controller. By means of the linear matrix inequality, we have successfully converted the controller existence problem into the feasibility of a set of matrix inequalities over a finite horizon. We should point out that the randomly occurring sensor failure model considered here is quite general, and therefore can be applied in many branches of fault-tolerant control and signal processing problems, such as control with actuator saturation, filtering with quantization effects and so on.*

Computational Algorithm

It can be noticed that the feedback controller gains can be obtained by solving the presented LMIs iteratively. In the next stage, it is our aim to propose a computational algorithm to get the numerical values of $A_g(k)$, $B_g(k)$ and $C_g(k)$ at the corresponding time point k recursively.

Reliable $\mathcal{H}_\infty$ Controller Design Algorithm

Step 1   Given the $\mathcal{H}_\infty$ performance γ, the positive weighted matrix W and the initial condition $x(0)$, select the initial values for matrices $\{P_1(0), P_2(0), P_3(0)\}$ satisfying condition (5.34). Set $k = 0$ and the final time N.

Step 2   Obtain the values of matrices $\{Q_1(k+1), Q_2(k+1), Q_3(k+1)\}$ and the desired controller parameters $\{A_g(k), B_g(k), C_g(k)\}$ at time point k by solving the LMIs (5.32)–(5.33).

Step 3   Set $k = k+1$ and obtain $\{P_1(k+1), P_2(k+1), P_3(k+1)\}$ using the parameter update formula (5.35).

Step 4   If $k < N$, then go to Step 2, or else stop the iteration.

Remark 5.5 *The Reliable $\mathcal{H}_\infty$ Controller Design Algorithm gives a recursive way to obtain the numerical values of the feedback gains at each instant k. It should be noticed that the existence of the controller is expressed by the feasibility of certain LMIs which could be solved iteratively forward in time. It therefore is suitable for the online design. A possible research topic in the future is to take more performance indices into consideration and design a multiobjective controller.*

5.4   Numerical Example

In this section, a numerical example is presented to demonstrate the effectiveness of the method proposed in this chapter.

Set $N = 50$, $\gamma = 1.5$, $W = 0.2I$ and the initial value $x(0) = 0$, $P_1(0) = I$ and $P_2(0) = P_3(0) = 0$. Consider a nonlinear stochastic system with the following parameters:

$$
\begin{aligned}
A(k) &= \begin{bmatrix} -0.15 & 0.03 \\ -0.2 + 0.2\sin(2k) & -0.1 \end{bmatrix}, \\
B(k) &= \begin{bmatrix} -0.1 \\ -0.5 + 0.05\cos(5k) \end{bmatrix}, \\
D(k) &= \begin{bmatrix} 0.1\sin(0.2k) \\ -0.02 \end{bmatrix}, \\
C(k) &= \begin{bmatrix} -0.1 & -0.1\sin(1.5k) - 0.25 \end{bmatrix}, \\
\Gamma_i(k) &= \begin{bmatrix} 0.09 & 0 \\ 0 & 0.16 \end{bmatrix}, \qquad E(k) = 0.1, \\
\pi_{1i}(k) &= \begin{bmatrix} 0.2 \\ 0.3 \end{bmatrix}, \qquad \pi_{2i}(k) = 0.5.
\end{aligned}
\tag{5.41}
$$

Assume $\mathcal{P}\{\alpha(k) = 1\} = \mathbb{E}\{\alpha(k)\} = \bar{\alpha} = 0.5$; then we can easily have $\sigma_\alpha^2 = 0.25$. Let the sensor failure matrix be $F(x(k), k) = 1.2\sin^2(x(k))$. Therefore, it can be obtained that $0 \leq F(x(k), k) \leq 1.2$ and $H(k) = 0.6$, $L(k) = 1$. Using the developed computational algorithm, we can solve the corresponding LMIs and then obtain the desired controller parameters, some of which are shown in Table 5.1.

TABLE 5.1: The output feedback controller parameters

time	$k = 0$		$k = 1$		$k = 2$		$\cdots$
$A_g(k)$	$\begin{bmatrix} 0.045 & -0.07 \\ -0.13 & 0.06 \end{bmatrix}$		$\begin{bmatrix} -0.05 & -0.28 \\ -0.35 & 0.05 \end{bmatrix}$		$\begin{bmatrix} -0.03 & -0.27 \\ -0.33 & 0.06 \end{bmatrix}$		$\cdots$
$B_g(k)$	$\begin{bmatrix} 1.18 & 0.88 \end{bmatrix}^T$		$\begin{bmatrix} 1.01 & 0.82 \end{bmatrix}^T$		$\begin{bmatrix} 1.06 & 0.86 \end{bmatrix}^T$		$\cdots$
$C_g(k)$	$\begin{bmatrix} 0.01 & -1.22 \end{bmatrix}$		$\begin{bmatrix} 0 & -1.56 \end{bmatrix}$		$\begin{bmatrix} 0 & -1.60 \end{bmatrix}$		$\cdots$

5.5   Summary

In this chapter, the reliable $\mathcal{H}_\infty$ control problem has been studied for a type of nonlinear stochastic systems with sensor failures. The stochastic nonlinearities taken into consideration could cover several well-studied nonlinearities. The sensor failures occur in a random way, and the failure probability is described by a stochastic variable satisfying the Bernoulli distribution. The solvability of the addressed control problem is expressed by the feasibility of certain linear matrix inequalities. The numerical values of the controller gains can be obtained by the given computing algorithm. An illustrative example is given to show the effectiveness and applicability of the proposed design strategy.

6

Event-Triggered Mean Square Consensus Control for Time-Varying Stochastic Multi-Agent System with Sensor Saturations

The past decade has witnessed an ever-growing interest in the study of so-called multi-agent systems (MASs) that have found extensive applications in various areas including unmanned aerial vehicles (UAVs), autonomous underwater vehicles (AUVs), automated highway systems (AHSs) and mobile robotics. Among popular research issues regarding MASs, the so-called consensus problem is concerned with the process where a batch of interacting agents governed by certain interconnection topology achieve a collective goal (e.g. the same trajectory). Owing to their clear engineering insights, the consensus behaviors of MASs have attracted a surge of research attention leading to a rich body of literature.

Up to now, most MASs discussed in the literature have been assumed to be *time-invariant*. This assumption is, however, very restrictive as almost all real-world engineering systems have certain parameters/structures which are indeed *time-varying*. For such time-varying systems, a finite-horizon controller is usually desirable as it could provide better transient performance for the controlled system especially when the noise inputs are non-stationary. However, when it comes to the consensus of multi-agent systems, the corresponding results have been scattered due mainly to the difficulty of quantifying the consensus over a finite horizon. It is notable that the consensus problem for MASs with time-varying parameters has received some initial research attention. Nevertheless, the research on time-varying multi-agent systems is far from adequate and there are still many open challenging problems remaining for further investigation. On the other hand, sensor saturation is a frequently encountered phenomenon resulting from physical limitations of system components as well as the difficulties of ensuring high fidelity and timely arrival of the control and sensing signals through a possibly unreliable network of limited bandwidth. In other words, the sensor outputs are often saturated because the physical entities or processes cannot transmit energy and power with unbounded magnitude or rate. As such, it makes practical sense to take sensor saturation into account when dealing with the output-feedback control problems for time-varying MASs, which remains as an ongoing research issue.

On another research frontier, event-triggered control/filtering strategies have recently become an attractive area of research because of their capabilities in improving the resource utilization efficiency by reducing the unnecessary executions as compared to the traditional time-triggered mechanism. In the context of MASs, so far, much work has been done for event-triggered consensus control and most available results have been restricted to linear time-invariant MASs only. When it comes to the time-varying MASs, the corresponding event-triggered schemes have received very little research effort due probably to the technical difficulty in handling the time-varying coupling between the triggering mechanism and interaction topology in case of communications among the agents. Up to now, the mean square consensus control problem for *time-varying* MASs with event-triggered mechanism has not been adequately investigated, not to mention the case where sensor saturations are also involved. Such a situation has motivated the present investigation.

In this chapter, we endeavor to design an event-triggered output-feedback controller for a class of discrete time-varying stochastic MAS to reach a new kind of mean square consensus with guaranteed upper bound on the consensus index subject to sensor saturation. In doing so, three technical challenges are identified as follows: 1) how to define the consensus of MASs in a time-varying context; 2) how to develop appropriate analysis and synthesis techniques associated with time-varying MASs; and 3) how to establish a unified framework to handle the cross coupling among topology, time-varying parameters, event-triggered mechanism as well as the sensor saturations. In fact, the main purpose of this chapter is to provide satisfactory answers to these questions by launching a major study.

The rest of this chapter is organized as follows. Section 6.1 formulates the event-triggered output-feedback consensus control problem for discrete time-varying stochastic MAS subject to sensor saturations. The main results are presented in Section 6.2 where sufficient conditions for the MAS to reach the mean square consensus with guaranteed performance are given in terms of recursive matrix inequalities. Section 6.3 gives a numerical example and Section 6.4 draws our conclusion.

6.1   Problem Formulation

In this chapter, the multi-agent system has N agents which communicate with each other according to a fixed network topology represented by an undirected graph $\mathscr{G} = (\mathscr{V}, \mathscr{E}, \mathscr{H})$ of order N with the set of agents $\mathscr{V} = \{1, 2, \ldots, N\}$, the set of edges $\mathscr{E} \in \mathscr{V} \times \mathscr{V}$, and the weighted adjacency matrix $\mathscr{H} = [h_{ij}]$ with nonnegative adjacency element h_{ij}. If $(i, j) \in \mathscr{E}$, then $h_{ij} > 0$, or else $h_{ij} = 0$. An edge of $\mathscr{G}$ is denoted by the ordered pair (i, j). The adjacency elements associated with the edges of the graph are positive; i.e., $h_{ij} > 0 \iff (i, j) \in \mathscr{E}$, which means that agent i can obtain information from agent j. Furthermore, self-edges (i, i) are not allowed, i.e., $(i, i) \notin \mathscr{E}$ for any $i \in \mathscr{V}$. The neighborhood of agent i is denoted by $\mathscr{N}_i = \{j \in \mathscr{V} : (j, i) \in \mathscr{E}\}$. The in-degree of agent i is defined as $\deg_{\text{in}}^i \triangleq \sum_{j \in \mathscr{N}_i} h_{ij}$.

Consider a discrete time-varying stochastic multi-agent system described by the following state-space model:

$$
\begin{align}
x_{i,k+1} &= A_k x_{i,k} + B_k u_{i,k} + D_k w_{i,k} \tag{6.1} \\
y_{i,k} &= \kappa(C_k x_{i,k}) + E_k v_{i,k} \tag{6.2}
\end{align}
$$

where $x_{i,k} \in \mathbb{R}^n$, $y_{i,k} \in \mathbb{R}^q$ and $u_{i,k} \in \mathbb{R}^p$ are, respectively, the state vector, the measurement output and the control input of agent i. A_k, B_k, C_k, D_k and E_k are time-varying matrices with compatible dimensions. $w_{i,k} \in \mathbb{R}^\omega$ and $v_{i,k} \in \mathbb{R}^\nu$ $(i = 1, 2, \ldots, N)$ are mutually uncorrelated zero-mean Gaussian white noise sequences. Denote $\tilde{w}_k \triangleq [w_{1,k}^{\text{T}} \cdots w_{N,k}^{\text{T}} \; v_{1,k}^{\text{T}} \cdots v_{N,k}^{\text{T}}]^{\text{T}}$. The statistical properties of $w_{i,k}$ and $v_{i,k}$ can be described as follows:

$$
\mathbb{E}\{\tilde{w}_k\} = 0,
$$

$$
\mathbb{E}\{\tilde{w}_k \tilde{w}_l^{\text{T}}\} = \begin{bmatrix} \text{diag}_N\{W_{i,k}\delta_{kl}\} & 0 \\ 0 & \text{diag}_N\{V_{i,k}\delta_{kl}\} \end{bmatrix}
$$

where $W_{i,k}$ and $V_{i,k}$ $(i = 1, 2, \ldots, N)$ are all known positive definite matrices, and δ_{kl} is defined by:

$$
\delta_{kl} = \begin{cases} I & k = l \\ 0 & k \neq l \,. \end{cases}
$$

The saturation function $\kappa(\cdot)$ in (6.2) is defined as

$$
\kappa(r) \triangleq \text{col}_q\{\kappa_i(r^{(i)})\} \tag{6.3}
$$

where $\kappa_i(r^{(i)}) = \text{sign}(r^{(i)}) \min\{r_{\max}^{(i)}, |r^{(i)}|\}$ with $r^{(i)}$ denoting the ith entry of the vector r.

Definition 6.1 *Let U_1 and U_2 be real matrices with $U \triangleq U_2 - U_1 > 0$. A nonlinearity $\varphi(\cdot)$ is said to satisfy the sector condition with respect to U_1 and U_2 if*

$$
\left(\varphi(y) - U_1 y\right)^{\text{T}} \left(\varphi(y) - U_2 y\right) \leq 0. \tag{6.4}
$$

In this case, the sector-bounded nonlinearity $\varphi(\cdot)$ is said to belong to the sector $[U_1, U_2]$.

Noting that if there exist diagonal matrices G_1 and G_2 such that $0 \leq G_1 < I \leq G_2$, then the saturation function $\kappa(C_k x_{i,k})$ in (6.2) can be written as follows:

$$
\kappa(C_k x_{i,k}) = G_1 C_k x_{i,k} + \varphi(C_k x_{i,k}) \tag{6.5}
$$

where $\varphi(C_k x_{i,k})$ is a nonlinear vector-valued function satisfying the sector condition with $U_1 = 0$ and $U_2 = G \triangleq G_2 - G_1$; i.e., $\varphi(C_k x_{i,k})$ satisfies the following inequality:

$$\varphi^{\mathrm{T}}(C_k x_{i,k})\big(\varphi(C_k x_{i,k}) - GC_k x_{i,k}\big) \le 0. \tag{6.6}$$

In this chapter, a control protocol of the following form is adopted:

$$u_{i,k} = K_k \eta_{i,k} \quad \text{with} \quad \eta_{i,k} = \sum_{j \in \mathcal{N}_i} h_{ij}\big(y_{j,k} - y_{i,k}\big) \tag{6.7}$$

where K_k is the feedback gain to be designed and $\eta_{i,k}$ represents the updating signal feeding to the controller of agent i.

Let us now discuss the event-triggering mechanism to be adopted. Suppose that the sequence of the triggering instants is $\{k_t^i\}$ $(t = 0, 1, 2, \ldots)$ satisfying $0 < k_0^i < k_1^i < k_2^i < \cdots < k_t^i < \cdots$, where k_t^i represents the time instant k when the $(t+1)$-th trigger occurs for agent i. Then, for $k > k_t^i$, define

$$e_{i,k} \triangleq \eta_{i,k_t^i} - \eta_{i,k} \tag{6.8}$$

with η_{i,k_t^i} representing the updating signal feeding to the controller of agent i at the latest triggering time k_t^i. Then, the sequence of event-triggering instants is determined iteratively by

$$k_{t+1}^i = \inf\{k \in \mathbb{Z}^+ | k > k_t^i,\ e_{i,k}^{\mathrm{T}}\Omega_{i,k}^{-1}e_{i,k} > 1\} \tag{6.9}$$

where $\Omega_{i,k} > 0$ is referred to as the triggering threshold matrix.

Applying the event-triggering mechanism, we can rewrite the updating signal $\eta_{i,k}$ defined in (6.7) as follows:

$$\eta_{i,k} = \begin{cases} 0, & k \in [0, k_0^i) \\ \eta_{i,k_t^i}, & k \in [k_t^i, k_{t+1}^i) \end{cases} \tag{6.10}$$

which implies that the controller input defined in (6.7) remains a constant in the execution interval $[k_t, k_{t+1})$. Without loss of generality, it is assumed that $\eta_{i,k_t^i} = 0$ when $k \in [0\ k_0^i)$. The control law can be now rewritten as

$$u_{i,k} = K_k(\eta_{i,k} + e_{i,k}), \quad e_{i,k}^{\mathrm{T}}\Omega_{i,k}^{-1}e_{i,k} \le 1. \tag{6.11}$$

Implementing control law (6.11) to MAS (6.1)–(6.2), we obtain the following closed-loop system:

$$\begin{aligned} x_{k+1} =& \big(I_N \otimes A_k + \mathcal{H}_k \otimes (B_k K_k G_1 C_k)\big)x_k \\ &+ (I_N \otimes D_k)w_k + \big(\mathcal{H}_k \otimes (B_k K_k E_k)\big)v_k \\ &+ \big(\mathcal{H}_k \otimes (B_k K_k)\big)\varphi_k + \big(I_N \otimes (B_k K_k)\big)e_k \end{aligned} \tag{6.12}$$

where
$$x_k = \text{col}_N\{x_{i,k}\},$$
$$e_k = \text{col}_N\{e_{i,k}\},$$
$$\varphi_k = \text{col}_N\{\kappa(C_k x_{i,k})\},$$
$$w_k = \text{col}_N\{w_{i,k}\},$$
$$v_k = \text{col}_N\{v_{i,k}\},$$
$$\mathcal{H}_k = \begin{bmatrix} -\deg_{\text{in}}^1 & h_{1,2} & h_{1,3} & \cdots & h_{1,N} \\ h_{2,1} & -\deg_{\text{in}}^2 & h_{2,3} & \cdots & h_{2,N} \\ \vdots & \vdots & \vdots & \ddots & \vdots \\ h_{N,1} & h_{N,2} & h_{N,3} & \cdots & -\deg_{\text{in}}^N \end{bmatrix}.$$

In order to discuss the consensus performance of MAS (6.1)–(6.2) in the mean square, we first denote the average state of all agents by:

$$\bar{x}_k \triangleq \mathbb{E}\left\{ \frac{1}{N} \sum_{i=1}^N x_{i,k} \Big| y_{k-1} \right\} = \frac{1}{N} \mathbb{E}\left\{ (\mathbf{1}_N^{\mathrm{T}} \otimes I_n) x_k \Big| y_{k-1} \right\} \tag{6.13}$$

where $y_{k-1} \triangleq \text{col}_N\{y_{i,k-1}\}$ represents the measurements obtained at time instant $k - 1$. Then, at time instant k, given the measurements $y_k \triangleq \text{col}_N\{y_{i,k}\}$ (which means that e_k is also available according to (6.7)–(6.8)), we can calculate $\bar{x}_{k+1}$ by

$$\bar{x}_{k+1} = \frac{1}{N} \mathbb{E}\left\{ (\mathbf{1}_N^{\mathrm{T}} \otimes I_n) x_{k+1} \Big| y_k \right\}$$
$$= A_k \bar{x}_k + \frac{1}{N} (\mathbf{1}_N^{\mathrm{T}} \otimes (B_k K_k)) e_k. \tag{6.14}$$

It should be pointed out that recursion (6.14) plays a pivotal role in computing the expected average state at each time instant and subsequently in obtaining the desired feedback gain. Such a procedure will be discussed later in more detail.

Definition 6.2 *The performance index of the mean square consensus for agent i ($i = 1, 2, \ldots, N$) of the time-varying stochastic multi-agent system (6.1)–(6.2) at time instant k is defined by*

$$\mathfrak{D}_{i,k} \triangleq \mathbb{E}\left\{ (x_{i,k} - \bar{x}_k)(x_{i,k} - \bar{x}_k)^{\mathrm{T}} \right\}. \tag{6.15}$$

Remark 6.1 *The performance index $\mathfrak{D}_{i,k}$ of the mean square consensus characterizes the deviation level from the agent i to the expected average of the states $\bar{x}_k$ at time instant k, thereby reflecting the transient consensus accuracy during the dynamical consensus process. Such an index, which can be intuitively understood as the "distance" from agent i to the expected center of the MAS (characterized by $\bar{x}_k$) at time step k, is proposed in response to the* consideration of the additive noises $w_{i,k}$ and $v_{i,k}$, the event-triggered mechanism (6.9) as well as the time-varying nature of MAS (6.1)–(6.2). In general, a smaller $\mathfrak{D}_{i,k}$ (in the sense of matrix trace) is indicative of a better consensus performance at time instant k.

Assumption 6.1 *The initial values of each agent, namely, $x_{i,0}$ ($i = 1, 2, \ldots, N$) are known and satisfy*

$$(x_{i,0} - \bar{x}_0)(x_{i,0} - \bar{x}_0)^{\mathrm{T}} \leq \Gamma_0 \tag{6.16}$$

where $\Gamma_0 > 0$ is a known positive definite matrix.

Definition 6.3 *Let the undirected communication graph $\mathscr{G}$, a sequence of triggering threshold matrices $\{\Omega_{i,k}\}_{k\geq 0}$ and a sequence of positive definite matrices $\{\Gamma_k\}_{k\geq 0}$ be given. The MAS (6.1)-(6.2) is said to reach mean square consensus with respect to the triple $(\mathscr{G}, \{\Omega_{i,k}\}, \{\Gamma_k\})$ if*

$$\mathfrak{D}_{i,k} \leq \Gamma_k, \qquad \forall i \in \mathscr{V}, \ k \geq 0 \tag{6.17}$$

hold at each time instant k, where $\mathfrak{D}_{i,k}$ is defined in (6.15).

Our objective of this chapter is twofold. First, we aim to design the sequence of output-feedback gains $\{K_k\}_{k\geq 0}$ such that MAS (6.1)-(6.2) reaches mean square consensus with respect to $(\mathscr{G}, \{\Omega_{i,k}\}, \{\Gamma_k\})$. Second, we aim to solve two optimization problems which, respectively, minimize Γ_k (in the sense of matrix trace) to seek the locally best consensus performance and maximize $\Omega_{i,k}$ (in the sense of matrix trace) to design the locally lowest triggering frequency at each time instant.

6.2   Main Results

Lemma 6.1 *A symmetric matrix $P \in \mathbb{R}^{\epsilon \times \epsilon}$ is positive definite if and only if there exist $\rho_l \in \mathbb{R}^{\epsilon}$ ($l = 1, 2, \ldots, \epsilon$) such that $P = \sum_{l=1}^{\epsilon} \rho_l \rho_l^{\mathrm{T}}$ and $\mathrm{rank}[\rho_1 \ \rho_2 \ \cdots \ \rho_\epsilon] = \epsilon$.*

Lemma 6.2 *(S-procedure) Let $\psi_0(\cdot)$, $\psi_1(\cdot)$, ..., $\psi_m(\cdot)$ be quadratic functions of the variable $\varsigma \in \mathbb{R}^n$: $\psi_j(\varsigma) \triangleq \varsigma^{\mathrm{T}} T_j \varsigma$ ($j = 0, \ldots, m$), where $T_j^{\mathrm{T}} = T_j$. If there exist $\tau_1 \geq 0$, ..., $\tau_m \geq 0$ such that $\varsigma^{\mathrm{T}}(T_0 - \sum_{j=1}^{m} \tau_j T_j)\varsigma \leq 0$, then the following is true:*

$$\psi_1(\varsigma) \leq 0, \ldots, \psi_m(\varsigma) \leq 0 \rightarrow \psi_0(\varsigma) \leq 0. \tag{6.18}$$

Lemma 6.3 *(Schur Complement Lemma) Given constant matrices $\mathcal{S}_1, \mathcal{S}_2, \mathcal{S}_3$ where $\mathcal{S}_1 = \mathcal{S}_1^{\mathrm{T}}$ and $0 < \mathcal{S}_2 = \mathcal{S}_2^{\mathrm{T}}$, then $\mathcal{S}_1 + \mathcal{S}_3^{\mathrm{T}} \mathcal{S}_2^{-1} \mathcal{S}_3 < 0$ if and only if*

$$\begin{bmatrix} \mathcal{S}_1 & \mathcal{S}_3^{\mathrm{T}} \\ \mathcal{S}_3 & -\mathcal{S}_2 \end{bmatrix} < 0, \quad \text{or} \quad \begin{bmatrix} -\mathcal{S}_2 & \mathcal{S}_3 \\ \mathcal{S}_3^{\mathrm{T}} & \mathcal{S}_1 \end{bmatrix} < 0. \tag{6.19}$$

Consensus control subject to a fixed triple $(\mathscr{G}, \Omega_{i,k}, \Gamma_k)$

For simplicity of the following notation, we denote

$$\mathcal{W}_k \triangleq \operatorname{diag}_N\{W_i\},$$

$$\mathcal{V}_k \triangleq \operatorname{diag}_N\{V_i\},$$

$$\mathcal{D}_k \triangleq I_N \otimes D_k,$$

$$\mathcal{E}_k \triangleq \mathcal{H}_k \otimes (B_k K_k E_k),$$

$$\mathscr{D}_k \triangleq \mathcal{D}_k^{\mathrm{T}} \mathcal{L}_{n,i}^{\mathrm{T}} \Gamma_{k+1}^{-1} \mathcal{L}_{n,i} \mathcal{D}_k \mathcal{W}_k,$$

$$\mathscr{E}_k \triangleq \mathcal{E}_k^{\mathrm{T}} \mathcal{L}_{n,i}^{\mathrm{T}} \Gamma_{k+1}^{-1} \mathcal{L}_{n,i} \mathcal{E}_k \mathcal{V}_k,$$

$$\mathcal{T}_{1,k} \triangleq \operatorname{diag}\{-1, \mathcal{L}_{n,i}^{\mathrm{T}} \mathcal{L}_{n,i}, 0, 0\},$$

$$\mathcal{T}_{2,k} \triangleq \operatorname{diag}\left\{-1, 0, \mathcal{L}_{q,i}^{\mathrm{T}} \Omega_{i,k}^{-1} \mathcal{L}_{q,i}, 0\right\},$$

$$\bar{\Pi}_{12} \triangleq I_N \otimes (A_k F_k) + \mathcal{H}_k \otimes (B_k K_k G_1 C_k F_k),$$

$$\bar{\Pi}_k \triangleq \begin{bmatrix} 0 & \bar{\Pi}_{12} & -\mathcal{N}_k \otimes (B_k K_k) & \mathcal{H}_k \otimes (B_k K_k) \end{bmatrix},$$

$$\bar{\Psi}_k \triangleq \begin{bmatrix} -(1_N \otimes (GC_k))\bar{x}_k & -I_N \otimes (GC_k F_k) & 0 \end{bmatrix},$$

$$\Psi_k \triangleq \frac{1}{2} \begin{bmatrix} 0 & \bar{\Psi}_k^{\mathrm{T}} \\ \bar{\Psi}_k & 2I_{nN} \end{bmatrix},$$

$$\mathcal{L}_{\sigma,i} \triangleq \begin{bmatrix} \underbrace{0 \;\cdots\; 0}_{i-1} & I_\sigma & \underbrace{0 \;\cdots\; 0}_{N-i} \end{bmatrix}, \quad (\sigma = \{n, q\})$$

$$\mathcal{N}_k \triangleq [a_{ij}]_{N \times N} \text{ with } a_{ij} \triangleq \begin{cases} (1-N)/N, & i = j \\ -1/N, & i \neq j \end{cases}$$

Moreover, by Lemma 6.1, the matrices $\mathcal{W}_k$ and $\mathcal{V}_k$ can be decomposed by $\mathcal{W}_k = \sum_{l=1}^{\epsilon} \vartheta_{l,k}\vartheta_{l,k}^{\mathrm{T}}$ and $\mathcal{V}_k = \sum_{l=1}^{\varepsilon} \pi_{l,k}\pi_{l,k}^{\mathrm{T}}$ with $\vartheta_{l,k} \in \mathbb{R}^{\epsilon}$ and $\pi_{l,k} \in \mathbb{R}^{\varepsilon}$ ($\epsilon = N\omega$, $\varepsilon = N\nu$).

Theorem 6.1 *Let the triple* $(\mathscr{G}, \{\Omega_{i,k}\}, \{\Gamma_k\})$ *be given. MAS* (6.1)–(6.2) *reaches mean square consensus with respect to the triple* $(\mathscr{G}, \{\Omega_{i,k}\}, \{\Gamma_k\})$ *if there exist a sequence of real-valued matrices* $\{K_k\}_{k\geq 0}$, *sequences of positive scalars* $\{\gamma_{i,k}\}_{k\geq 0}$ *and* $\{\lambda_{i,k}\}_{k\geq 0}$, *sequences of non-negative scalars* $\{\tau_{i,k}^{(1)}\}_{k\geq 0}$, $\{\tau_{i,k}^{(2)}\}_{k\geq 0}$ *and* $\{\tau_k^{(3)}\}_{k\geq 0}$ $(i = 1, 2, \ldots N)$ *such that the following recursive linear matrix inequalities (RLMIs) are true:*

$$\begin{bmatrix} -\gamma_{i,k} & \bar{\vartheta}_k^{\mathrm{T}} \\ \bar{\vartheta}_k & -\operatorname{diag}_\epsilon\{\Gamma_{k+1}\} \end{bmatrix} \leq 0, \tag{6.20}$$

$$\begin{bmatrix} -\lambda_{i,k} & \bar{\pi}_k^{\mathrm{T}} \\ \bar{\pi}_k & -\operatorname{diag}_\varepsilon\{\Gamma_{k+1}\} \end{bmatrix} \leq 0, \tag{6.21}$$

$$\begin{bmatrix} -\tilde{\Theta}_k & \bar{\Pi}_k^{\mathrm{T}} \mathcal{L}_{n,i}^{\mathrm{T}} \\ \mathcal{L}_{n,i}\bar{\Pi}_k & -\Gamma_{k+1} \end{bmatrix} \leq 0 \tag{6.22}$$

where

$$\bar{\vartheta}_k = \big(I_\epsilon \otimes (\mathcal{L}_{n,i}\mathcal{D}_k)\big)\mathrm{col}_\epsilon\{\vartheta_{i,k}\},$$

$$\bar{\pi}_k = \big(I_\varepsilon \otimes (\mathcal{L}_{n,i}\mathcal{E}_k)\big)\mathrm{col}_\varepsilon\{\pi_{i,k}\},$$

$$\tilde{\Theta}_1 = 1 - (\gamma_{i,k} + \lambda_{i,k}) - \sum_{i=1}^{N}\left(\tau_{i,k}^{(1)} + \tau_{i,k}^{(2)}\right),$$

$$\tilde{\Theta}_2 = \sum_{i=1}^{N}\tau_{i,k}^{(1)}\mathcal{L}_{n,i}^{\mathrm{T}}\mathcal{L}_{n,i},$$

$$\tilde{\Theta}_3 = \sum_{i=1}^{N}\tau_{i,k}^{(2)}\mathcal{L}_{q,i}^{\mathrm{T}}\Omega_{i,k}^{-1}\mathcal{L}_{q,i},$$

$$\tilde{\Theta}_k = \tau_k^{(3)}\Psi_k + \mathrm{diag}\Big\{\tilde{\Theta}_1,\tilde{\Theta}_2,\tilde{\Theta}_3,0\Big\} \tag{6.23}$$

with $F_k \in \mathbb{R}^{n \times n}$ being a factorization of Γ_k (i.e., $\Gamma_k = F_k F_k^{\mathrm{T}}$).

Proof *First of all, denote $\tilde{x}_{i,k} \triangleq x_{i,k} - \bar{x}_k$ and $\tilde{x}_k \triangleq \mathrm{col}_N\{\tilde{x}_{i,k}\}$. By subtracting (6.14) from (6.1), we obtain*

$$\begin{aligned}
\tilde{x}_{i,k+1} &= x_{i,k+1} - \bar{x}_{k+1} \\
&= A_k x_{i,k} + B_k\big(K_k(\eta_{i,k} + e_{i,k})\big) + D_k w_{i,k} \\
&\quad - \Big(A_k\bar{x}_k + \frac{1}{N}\big(\mathbf{1}_N^{\mathrm{T}} \otimes (B_k K_k)\big)e_k\Big). \tag{6.24}
\end{aligned}$$

Taking (6.12) into consideration, we can easily acquire

$$\begin{aligned}
\tilde{x}_{k+1} &= x_{k+1} - (\mathbf{1}_N \otimes I_n)\bar{x}_{k+1} \\
&= \big(I_N \otimes A_k + \mathcal{H}_k \otimes (B_k K_k G_1 C_k)\big)x_k \\
&\quad + \big(\mathcal{H}_k \otimes (B_k K_k)\big)\varphi_k + (I_N \otimes D_k)w_k \\
&\quad + \big(\mathcal{H}_k \otimes (B_k K_k E_k)\big)v_k + \big(I_N \otimes (B_k K_k)\big)e_k \\
&\quad - (\mathbf{1}_N \otimes I_n)A_k\bar{x}_k - \frac{1}{N}\big((\mathbf{1}_N\mathbf{1}_N^{\mathrm{T}}) \otimes (B_k K_k)\big)e_k \\
&= \big(I_N \otimes A_k + \mathcal{H}_k \otimes (B_k K_k G_1 C_k)\big)x_k \\
&\quad + \mathcal{D}_k w_k + \mathcal{E}_k v_k + \big(\mathcal{H}_k \otimes (B_k K_k)\big)\varphi_k \\
&\quad - (\mathbf{1}_N \otimes A_k)\bar{x}_k - \big(\mathcal{N}_k \otimes (B_k K_k)\big)e_k. \tag{6.25}
\end{aligned}$$

The rest of the proof is performed by induction. It follows directly from (6.16) that, when $k = 0$, $\mathfrak{D}_{i,0} \le \Gamma_0$ ($\forall i \in \mathcal{V}$) is satisfied. Supposing that $\mathfrak{D}_{i,k} \le \Gamma_k$ holds at time instant k, then it remains to prove that $\mathfrak{D}_{i,k+1} \le \Gamma_{k+1}$ also holds with the condition given in the theorem. Next, it can be easily verified that if

$$\mathbb{E}\{(x_{i,k} - \bar{x}_k)(x_{i,k} - \bar{x}_k)^{\mathrm{T}}\} \le \Gamma_k, \tag{6.26}$$

then there exists $z_{i,k} \in \mathbb{R}^n$ with $\mathbb{E}\{z_{i,k}z_{i,k}^{\mathrm{T}}\} \le I_n$ such that

$$x_{i,k} = \bar{x}_k + F_k z_{i,k} \tag{6.27}$$

where $F_k \in \mathbb{R}^{n \times n}$ is a factorization of Γ_k (i.e., $\Gamma_k = F_k F_k^\mathrm{T}$). Hence, with $z_k \triangleq \mathrm{col}_N\{z_{i,k}\}$, it follows from (6.27) that

$$x_k = (\mathbf{1}_N \otimes I_n)\bar{x}_k + (I_N \otimes F_k)z_k. \tag{6.28}$$

With the help of (6.28), we obtain from (6.25) that

$$
\begin{aligned}
\tilde{x}_{k+1} =& x_{k+1} - (\mathbf{1}_N \otimes I_n)\bar{x}_{k+1} \\
=& \big(I_N \otimes A_k + \mathcal{H}_k \otimes (B_k K_k G_1 C_k)\big) \\
& \times \big((\mathbf{1}_N \otimes I_n)\bar{x}_k + (I_N \otimes F_k)z_k\big) \\
& + \big(\mathcal{H}_k \otimes (B_k K_k)\big)\varphi_k + (I_N \otimes D_k)w_k \\
& + \big(\mathcal{H}_k \otimes (B_k K_k E_k)\big)v_k - (\mathbf{1}_N \otimes A_k)\bar{x}_k \\
& - \big(\mathcal{N}_k \otimes (B_k K_k)\big)e_k \\
=& (\mathbf{1}_N \otimes A_k)\bar{x}_k + \big(I_N \otimes (A_k F_k)\big)z_k \\
& + \big(\mathcal{H}_k \otimes (B_k K_k G_1 C_k F_k)\big)z_k + (I_N \otimes D_k)w_k \\
& + \big(\mathcal{H}_k \otimes (B_k K_k)\big)\varphi_k + \big(\mathcal{H}_k \otimes (B_k K_k E_k)\big)v_k \\
& - (\mathbf{1}_N \otimes A_k)\bar{x}_k - \big(\mathcal{N}_k \otimes (B_k K_k)\big)e_k \\
=& (I_N \otimes D_k)w_k + \big(\mathcal{H}_k \otimes (B_k K_k E_k)\big)v_k \\
& + \big(I_N \otimes (A_k F_k) + \mathcal{H}_k \otimes (B_k K_k G_1 C_k F_k)\big)z_k \\
& - \big(\mathcal{N}_k \otimes (B_k K_k)\big)e_k + \big(\mathcal{H}_k \otimes (B_k K_k)\big)\varphi_k \\
=& \big(I_N \otimes (A_k F_k) + \mathcal{H}_k \otimes (B_k K_k G_1 C_k F_k)\big)z_k \\
& - \big(\mathcal{N}_k \otimes (B_k K_k)\big)e_k + \big(\mathcal{H}_k \otimes (B_k K_k)\big)\varphi_k \\
& + \mathcal{D}_k w_k + \mathcal{E}_k v_k. \tag{6.29}
\end{aligned}
$$

Subsequently, by denoting

$$
\xi_k \triangleq \begin{bmatrix} 1 \\ z_k \\ e_k \\ \varphi_k \end{bmatrix}, \tag{6.30}
$$

$$
\begin{aligned}
\Pi_k \triangleq \big[\; & \mathcal{D}_k w_k + \mathcal{E}_k v_k \;\; \bar{\Pi}_{12} \\
& -\mathcal{N}_k \otimes (B_k K_k) \;\; \mathcal{H}_k \otimes (B_k K_k) \; \big], \tag{6.31}
\end{aligned}
$$

we can further express $\tilde{x}_{k+1}$ in (6.29) as follows:

$$\tilde{x}_{k+1} = x_{k+1} - (\mathbf{1}_N \otimes I_n)\bar{x}_{k+1} = \Pi_k \xi_k. \tag{6.32}$$

Next, it follows from Lemma 6.3 that

$$
\begin{aligned}
& \mathbb{E}\{z_{i,k} z_{i,k}^\mathrm{T}\} \leq I_n \\
\Rightarrow & \mathbb{E}\{z_{i,k}^\mathrm{T} z_{i,k}\} \leq 1 \\
\Rightarrow & \mathbb{E}\{\xi_k^\mathrm{T} \mathcal{T}_{1,k} \xi_k\} \leq 0 \tag{6.33}
\end{aligned}
$$

where $T_{1,k}$ is defined previously.

By the same token, we can know from (6.11) that the vector $e_{i,k}$ satisfies

$$e_{i,k}^{\mathrm{T}}\Omega_{i,k}^{-1}e_{i,k} \leq 1, \tag{6.34}$$

which can be described by ξ_k as follows:

$$\xi_k^{\mathrm{T}}T_{2,k}\xi_k \leq 0 \tag{6.35}$$

where $T_{2,k}$ is defined previously.

Similarly, inequality (6.6) which characterizes the constraints resulting from the sensor saturations can be rewritten as

$$\big((I_N \otimes I_n)\varphi_k\big)^{\mathrm{T}}\Big(\varphi_k - \big(I_N \otimes (GC_k)\big)x_k\Big) \leq 0. \tag{6.36}$$

Substituting (6.27) into (6.36) leads to

$$\varphi_k^{\mathrm{T}}(I_N \otimes I_n)\Big(\varphi_k - \big(I_N \otimes (GC_k)\big)$$
$$\times \big((\mathbf{1}_N \otimes I_n)\bar{x}_k + (I_N \otimes F_k)z_k\big)\Big) \leq 0,$$

which can be equivalently expressed by ξ_k as

$$\xi_k^{\mathrm{T}}\Psi_k\xi_k \leq 0 \tag{6.37}$$

with Ψ_k being defined previously.

So far, in terms of the vector ξ_k, we have converted all the constraints imposed on the time-varying MAS (6.1)–(6.2) into certain inequalities (i.e. (6.33), (6.35) and (6.37)). It now remains to show that $\Gamma_{k+1} \leq 1$ holds if the condition of this theorem is satisfied at time instant k. To this end, by means of Lemma 6.3, the set of RLMIs (6.22) is feasible if and only if

$$-\tilde{\Theta}_k + \bar{\Pi}_k^{\mathrm{T}}\mathcal{L}_{n,i}^{\mathrm{T}}\Gamma_{k+1}^{-1}\mathcal{L}_{n,i}\bar{\Pi}_k \leq 0. \tag{6.38}$$

For brevity of later development, we denote

$$\vec{\Pi}_k \triangleq \bar{\Pi}_k^{\mathrm{T}}\mathcal{L}_{n,i}^{\mathrm{T}}\Gamma_{k+1}^{-1}\mathcal{L}_{n,i}\bar{\Pi}_k.$$

Substituting (6.23) into (6.38) yields

$$\xi_k^{\mathrm{T}}\Big(\vec{\Pi}_k - \tau_k^{(3)}\Psi_k - \mathrm{diag}\big\{\tilde{\Theta}_1, \tilde{\Theta}_2, \tilde{\Theta}_3, 0\big\}\Big)\xi_k \leq 0.$$

After some tedious but straightforward manipulations, we arrive at

$$\xi_k^{\mathrm{T}}\Big(\vec{\Pi}_k + \breve{\Pi}_k - \tau_k^{(3)}\Psi_k - \sum_{i=1}^{N}\big(\tau_{i,k}^{(1)}T_{1,k} + \tau_{i,k}^{(2)}T_{2,k}\big)\Big)\xi_k \leq 0 \tag{6.39}$$

where $\breve{\Pi}_k \triangleq \mathrm{diag}\{\gamma_{i,k} + \lambda_{i,k}, 0, 0, 0\} - \mathrm{diag}\{1, 0, 0, 0\}$.

By Lemma 6.2, it follows from (6.33), (6.35), (6.37) *and* (6.39) *that*

$$\xi_k^{\mathrm{T}}(\vec{\Pi}_k + \breve{\Pi}_k)\xi_k \leq 0. \tag{6.40}$$

According to Lemma 6.3, the set of RLMIs (6.20) *holds if and only if*

$$-\gamma_{i,k} + \sum_{l=1}^{\epsilon} \vartheta_{l,k}^{\mathrm{T}} \mathcal{D}_k^{\mathrm{T}} \mathcal{L}_{n,i}^{\mathrm{T}} \Gamma_{k+1}^{-1} \mathcal{L}_{n,i} \mathcal{D}_k \vartheta_{l,k} \leq 0 \tag{6.41}$$

which, by properties of matrix trace, are equivalent to

$$-\gamma_{i,k} + \sum_{l=1}^{\epsilon} \mathrm{tr}\left[\mathcal{D}_k^{\mathrm{T}} \mathcal{L}_{n,i}^{\mathrm{T}} \Gamma_{k+1}^{-1} \mathcal{L}_{n,i} \mathcal{D}_k \vartheta_{l,k} \vartheta_{l,k}^{\mathrm{T}}\right] \leq 0. \tag{6.42}$$

Since $\mathcal{W}_k = \sum_{l=1}^{\epsilon} \vartheta_{l,k} \vartheta_{l,k}^{\mathrm{T}}$, *inequalities* (6.42) *imply*

$$\mathrm{tr}\left[\mathscr{D}_k\right] = \mathrm{tr}\left[\mathcal{D}_k^{\mathrm{T}} \mathcal{L}_{n,i}^{\mathrm{T}} \Gamma_{k+1}^{-1} \mathcal{L}_{n,i} \mathcal{D}_k \mathcal{W}_k\right] \leq \gamma_{i,k}. \tag{6.43}$$

Along a similar line, it can be derived from RLMIs (6.21) *that*

$$\mathrm{tr}\left[\mathscr{E}_k\right] = \mathrm{tr}\left[\mathcal{E}_k^{\mathrm{T}} \mathcal{L}_{n,i}^{\mathrm{T}} \Gamma_{k+1}^{-1} \mathcal{L}_{n,i} \mathcal{E}_k \mathcal{V}_k\right] \leq \lambda_{i,k}. \tag{6.44}$$

Denoting $\tilde{\Pi}_k \triangleq [\mathcal{D}_k w_k + \mathcal{E}_k v_k \ 0 \ 0 \ 0]$ *and taking into account the statistical properties of random variables* w_k *and* v_k, *we obtain*

$$
\begin{aligned}
&\mathbb{E}\{\xi_k^{\mathrm{T}} \Pi_k^{\mathrm{T}} \mathcal{L}_{n,i}^{\mathrm{T}} \Gamma_{k+1}^{-1} \mathcal{L}_{n,i} \Pi_k \xi_k\} \\
=&\xi_k^{\mathrm{T}} \vec{\Pi}_k \xi_k + \mathbb{E}\{\xi_k^{\mathrm{T}} \tilde{\Pi}_k^{\mathrm{T}} \mathcal{L}_{n,i}^{\mathrm{T}} \Gamma_{k+1}^{-1} \mathcal{L}_{n,i} \tilde{\Pi}_k \xi_k\} \\
=&\xi_k^{\mathrm{T}} \vec{\Pi}_k \xi_k + \xi_k^{\mathrm{T}} \mathrm{diag}\{\mathrm{tr}[\mathscr{D}_k] + \mathrm{tr}[\mathscr{E}_k], 0, 0, 0\}\xi_k.
\end{aligned} \tag{6.45}
$$

Therefore, it can be verified from inequalities (6.43), (6.44) *and* (6.40) *that the following is true:*

$$\mathbb{E}\{\xi_k^{\mathrm{T}} \Pi_k^{\mathrm{T}} \mathcal{L}_{n,i}^{\mathrm{T}} \Gamma_{k+1}^{-1} \mathcal{L}_{n,i} \Pi_k \xi_k\} - 1 \leq 0. \tag{6.46}$$

Applying now Lemma 6.3 to inequalities (6.46), *we acquire*

$$\mathbb{E}\{\mathcal{L}_{n,i} \Pi_k \xi_k \xi_k^{\mathrm{T}} \Pi_k^{\mathrm{T}} \mathcal{L}_{n,i}^{\mathrm{T}}\} \leq \Gamma_{k+1} \tag{6.47}$$

which implies that $\mathfrak{D}_{i,k+1} \leq \Gamma_{k+1}$ $(i = 1, \ldots, N)$ *also hold and the induction is now accomplished. Consequently, MAS* (6.1)–(6.2) *reaches mean square consensus and the proof is thus complete.*

Remark 6.2 *By resorting to a set of recursive linear matrix inequalities, the solvability of the addressed mean square consensus problem is cast into the feasibility of a set of RLMIs. Thanks to the capability of the LMI approach [18], not only the existence condition but also the explicit value of the desired controller parameter at each time instant, if it exists, can be obtained via solving the corresponding set of LMIs. It should be pointed out that, however, there have been so far very few satisfactory results on the convergence of the*

RLMI-based algorithm. Actually, the algorithm relies largely on the system structure (the time-varying parameters and the topology information), the pre-specified constraints (the triggering thresholds, the performance upperbound and the noise intensities) as well as the selection of the initial values. If, at a certain time instant, the corresponding set of LMIs is infeasible and thus the value of the controller parameter cannot be found, we can, within the allowable range, change the initial values, adjust the triggering thresholds or relax the constraints on the performance index to obtain a feasible solution.

Remark 6.3 *Notice that the RLMI algorithm proposed in this chapter is based on LMI approach. As discussed in [18], the computational complexity of an LMI system is bounded by $O(\mathcal{P}\mathcal{Q}3\log(\mathcal{U}/\varepsilon))$ where $\mathcal{P}$ represents the row size, $\mathcal{Q}$ stands for the number of scalar decision variables, $\mathcal{U}$ is a data-dependent scaling factor and ε is the relative accuracy set for the algorithm. For instance, let us now look at the mean square consensus criterion for the MAS (6.1)–(6.2) (as proposed in Theorem 6.1), where the number of agents is N, the iteration time is $k_{\max}$ and the dimensions of variables are known from $x_{i,k} \in \mathbb{R}^n$, $y_{i,k} \in \mathbb{R}^q$, $u_{i,k} \in \mathbb{R}^p$, $w_{i,k} \in \mathbb{R}^\omega$ and $v_{i,k} \in \mathbb{R}^\nu$. The RLMI-based mean square consensus criterion is implemented recursively for $k_{\max}$ steps and, at each step, we need to solve $3N$ LMIs with $4Nn + 2Nq + 3N$ rows and $4N + pq + 1$ scalar variables. Accordingly, the computational complexity of the proposed RLMI algorithm can be represented by $O(4Nnpqk_{\max} + 2Npq^2k_{\max})$. We can now conclude that the computational complexity of our proposed RLMI algorithm depends linearly on the length of time interval $k_{\max}$ and the number of agents N. It is worth mentioning that the study of LMI optimization is very active in recent years within the communities of applied mathematics, control science and signal processing. We can expect substantial speedups in the near future.*

It follows from Theorem 6.1 that the desired control protocols could be a set if non-empty. An interesting issue would be to look for a certain optimal protocol among the feasible set based on some criteria of engineering significance. In the following, two optimization problems are discussed in order to seek the locally best consensus performance and locally lowest triggering frequency, respectively.

Optimization Problems

Problem 1: *Minimization of $\{\Gamma_k\}_{k>0}$ (in the sense of matrix trace) subject to fixed couple $(\mathcal{G}, \{\Omega_{i,k}\})$ for locally best consensus performance*

Corollary 6.1 *Let the pair $(\mathcal{G}, \{\Omega_{i,k}\})$ be given. A sequence of minimized $\{\Gamma_k\}_{k\geq0}$ (in the sense of matrix trace) can be obtained if there exist real-valued matrices $\{K_k\}_{k\geq0}$, positive scalars $\{\gamma_{i,k}\}_{k\geq0}$ and $\{\lambda_{i,k}\}_{k\geq0}$, non-negative scalars $\{\tau_{i,k}^{(1)}\}_{k\geq0}$, $\{\tau_{i,k}^{(2)}\}_{k\geq0}$ and $\{\tau_k^{(3)}\}_{k\geq0}$ $(i = 1, 2, \ldots N)$ solving the fol-*

lowing optimization problem:

$$\min_{\Gamma_{k+1}, K_k, \gamma_{i,k}, \lambda_{i,k}, \tau_{i,k}^{(1)}, \tau_{i,k}^{(2)}, \tau_k^{(3)}} \text{tr}[\Gamma_{k+1}] \tag{6.48}$$

$$\text{s.t. } (6.20) - (6.22).$$

Problem *2: Maximization of $\{\Omega_{i,k}\}_{k \geq 0}$ (in the sense of matrix trace) subject to fixed couple $(\mathscr{G}, \{\Gamma_k\})$ for locally lowest triggering frequency*

Corollary 6.2 *Let the pair $(\mathscr{G}, \Gamma_k)$ be given. A sequence of maximized $\Omega_{i,k}$ (in the sense of matrix trace) is guaranteed if there exist real-valued matrices $\{K_k\}_{k \geq 0}$ and $\{\Upsilon_{i,k}\}_{k \geq 0}$, positive scalars $\{\gamma_{i,k}\}_{k \geq 0}$ and $\{\lambda_{i,k}\}_{k \geq 0}$, nonnegative scalars $\{\tau_{i,k}^{(1)}\}_{k \geq 0}$, $\{\tau_{i,k}^{(2)}\}_{k \geq 0}$ and $\{\tau_k^{(3)}\}_{k \geq 0}$ $(i = 1, 2, \ldots, N)$ solving the following optimization problem:*

$$\min_{K_k, \Upsilon_{i,k}, \gamma_{i,k}, \lambda_{i,k}, \tau_{i,k}^{(1)}, \tau_{i,k}^{(2)}, \tau_k^{(3)}} \text{tr}\left[\sum_{i=1}^{N} \alpha_i \Upsilon_{i,k} \right] \tag{6.49}$$

$$\text{s.t. } (6.20) - (6.21) \ \& \ \begin{bmatrix} -\hat{\Theta}_k & * \\ \mathcal{L}_{n,i} \bar{\bar{\Pi}}_k & -\Gamma_{k+1} \end{bmatrix} \leq 0$$

where

$$\hat{\Theta}_k = \text{diag}\left\{ \tilde{\Theta}_1, \tilde{\Theta}_2, \sum_{i=1}^{N} \mathcal{L}_{q,i}^{\text{T}} \tau_{i,k}^{(2)} \Upsilon_{i,k} \mathcal{L}_{q,i}, 0 \right\} + \tau_k^{(3)} \Psi_k$$

and $\alpha_i > 0$ are weighting scalars satisfying $\sum_{i=1}^{N} \alpha_i = 1$. The triggering threshold matrix $\Omega_{i,k}$ can be computed by $\Omega_{i,k} = \Upsilon_{i,k}^{-1}$.

The proofs of Corollaries 6.1–6.2 are straightforward and thus omitted.

Remark 6.4 *Based on Theorem 6.1, Corollaries 6.1–6.2 convert the original optimization problems to certain eigenvalues problems (EVPs) for minimizing the sum of the eigenvalues (which is equivalent to the matrix trace) of certain variables subject to the RLMIs constraints. As discussed in [18], such EVPs can be solved numerically using the interior-point method efficiently. Moreover, note that the control law is a linear function of the output deviations. Therefore, the suboptimal control algorithms developed in Corollaries 6.1–6.2 can only be applied to the linear output feedback control case. The corresponding optimal consensus control problem via nonlinear feedback is one of the problems deserving our further investigation.*

Remark 6.5 *It should be pointed out that, within the proposed framework, the feedback gain K_k needs to be calculated first by using the global information on the topology $\mathcal{H}_k$ before the implementation. Then, according to the obtained control protocol, the agents will reach the desired consensus by using*

the neighbors' information only. In this consensus process, global informa-
tion is no longer required and, therefore, the multi-agent system works in a
distributed way. It is worth mentioning that there have been some research
papers coping with the MAS control problems by utilizing global information
(e. g., the topology structure or the maximum/minimum eigenvalues of Lapla-
cian matrices), see [244] for example. [39] is another quintessential example
where both centralized and distributed approaches have been developed to deal
with the event-triggered control for multi-agent systems. Nevertheless, it would
be interesting to develop a framework within which the control protocol can be
designed only using the neighbors' information about the agents, and this will
be one of our future research topics.

6.3   Illustrative Example

Consider a multi-agent system with following parameters:

$$A_k = \begin{bmatrix} 1 + 0.2\sin(0.3k) & 0.02 + 0.02\sin(k) \\ 0.02 & 1 + 0.2\sin(2k) \end{bmatrix},$$

$$B_k = \begin{bmatrix} 0.3 + 0.15\cos(3k) \\ 0.3 + 0.12e^{-k} \end{bmatrix},$$

$$D_k = \begin{bmatrix} 0.3 + 0.06\cos(3k) \\ 0.03 \end{bmatrix},$$

$$C_k = \begin{bmatrix} 0.2 + 0.03\sin(k) & 0.25 + 0.01\cos(4k) \end{bmatrix},$$

$$E_k = 0.3 + 0.03\sin(k),$$

$$W_{i,k} = V_{i,k} = 1,$$

$$G_1 = 0.9, \ G_2 = 1.$$

Let there be 4 agents connected according to an undirected graph $\mathcal{G}$ with
the associated matrix $\mathcal{H}$ set by

$$\mathcal{H} = \begin{bmatrix} -4 & 1 & 2 & 1 \\ 1 & -4 & 1 & 2 \\ 2 & 1 & -4 & 1 \\ 1 & 2 & 1 & -4 \end{bmatrix}.$$

In this simulation, the saturation threshold value $r_{\max} = 5$. The function
$\varphi(C_k x_{i,k})$ in (6.5) can then be described by $\varphi(C_k x_{i,k}) = \kappa(C_k x_{i,k}) - 0.9C_k x_{i,k}$.

Set the initial values of agents' states and the initial Γ_0 as follows:

$$x_{1,0} = \begin{bmatrix} 20 \\ 5 \end{bmatrix},$$

$$x_{2,0} = \begin{bmatrix} 25 \\ 15 \end{bmatrix},$$

$$x_{3,0} = \begin{bmatrix} 10 \\ 20 \end{bmatrix},$$

$$x_{4,0} = \begin{bmatrix} 5 \\ 30 \end{bmatrix},$$

$$\Gamma_0 = \begin{bmatrix} 147 & -75 \\ -75 & 226 \end{bmatrix}.$$

Then, it can be easily checked that the initial condition (6.16) is satisfied. By implementing the schemes proposed in Corollaries 6.1-6.2, the simulation results are shown in Table 6.1, Table 6.2 and Figs. 6.1–6.6.

Table I presents some of the output feedback controller gains by solving the RLMIs in Corollaries 6.1–6.2. It can be seen from Figs. 6.1–6.4 that the trajectories of each agent in *Problem 1* are much closer to the average state than those in *Problem 2*, which indicates that the algorithm proposed in Corollary 6.1 leads to a better consensus performance. As far as triggering frequency is concerned, the total triggering times are shown, with a comparison to the conventional time-based strategy, in Table 6.2 for both optimization problems. It can be observed that i) the proposed event-triggering mechanism can effectively reduce the triggering frequency; and ii) the total triggering times in *Problem 2* are less than those in *Problem 1* which implies, as we anticipate, that the triggering frequency can be further reduced if we implement the strategy provided in Corollary 6.2. Such a finding can be further verified via the comparison between Fig. 6.5 and Fig. 6.6.

Time	$k = 0$	$k = 1$	$k = 2$	$k = 3$	$k = 4$	$\cdots$
K_k (*OP1*)	0.35	0.28	0.16	0.25	0.36	$\cdots$
K_k (*OP2*)	0.27	0.46	0.35	0.57	0.22	$\cdots$

TABLE 6.1: The output-feedback controller gains at each time step

Methodology	Time-based method	*Problem 1*	*Problem 2*
Triggering times	180	150	112

TABLE 6.2: Comparison of triggering times

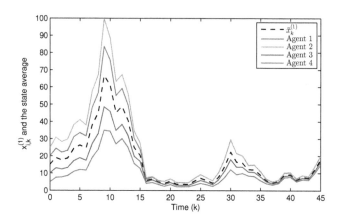

FIGURE 6.1: The trajectories of $x_{i,k}^{(1)}$ and the average $\bar{x}_k^{(1)}$ for *Problem 1*.

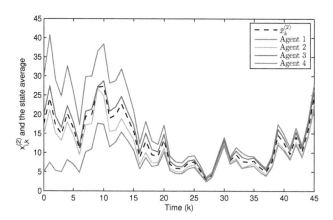

FIGURE 6.2: The trajectories of $x_{i,k}^{(2)}$ and the average $\bar{x}_k^{(2)}$ for *Problem 1*.

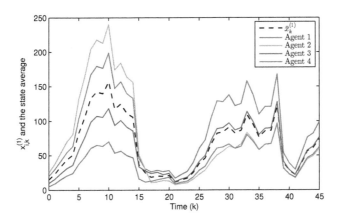

FIGURE 6.3: The trajectories of $x_{i,k}^{(1)}$ and the average $\bar{x}_k^{(1)}$ for *Problem 2*.

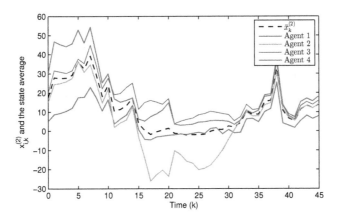

FIGURE 6.4: The trajectories of $x_{i,k}^{(2)}$ and the average $\bar{x}_k^{(2)}$ for *Problem 2*.

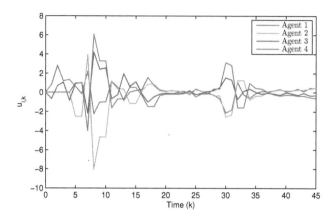

FIGURE 6.5: The control input $u_{i,k}$ for *Problem 1*.

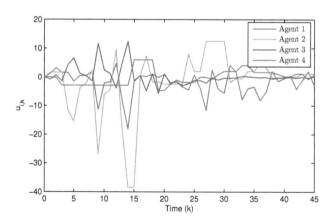

FIGURE 6.6: The control input $u_{i,k}$ for *Problem 2*.

6.4 Summary

In this chapter, the event-triggered mean square consensus control problem has been investigated for a class of discrete time-varying stochastic multi-agent systems subject to sensor saturations. First, a new definition of mean square consensus has been presented for the addressed MAS to characterize the transient consensus behavior. Then, by means of an RLMI approach, sufficient conditions have been established for the existence of the desired controller. Within the established framework, two optimization problems have been discussed to optimize the consensus performance and triggering frequency, respectively. Finally, an illustrative example has been exploited to show the effectiveness of the proposed control scheme.

7

Mean-Square $\mathcal{H}_\infty$ Consensus Control for A Class of Nonlinear Time-Varying Stochastic Multi-Agent Systems: The Finite-Horizon Case

The past decade has seen a surge of research interest in multi-agent systems (MASs) due primarily to their extensive applications in a variety of areas ranging from the chemistry manufacturing industry, geological exploration, and building automation, to military and aerospace industries. MASs consist of a large number of interaction/cooperation units called agents which directly interact with their neighbors according to a given topology. The MASs often display rich yet complex behavior even when all the agents have tractable models and interact with their neighbors in a simple and predictable fashion. Because of its clear physical and engineering insights, the consensus problem of MASs has been garnering considerable research attention and many results have been reported in the literature such as those on cooperative control of unmanned air vehicles (UAVs) or unmanned underwater vehicles (UUVs), formation control for multi-robot systems, collective behaviors of flocks or swarms and distributed sensor networks.

So far, almost all of the multi-agent systems investigated in the literature have been assumed to be *time-invariant*, and such an assumption is often not true for a large class of engineering systems whose parameters are indeed time-varying. For time-varying systems, a *finite-horizon* controller is usually desirable as it could provide better transient performance for the controlled system especially when the noise inputs are non-stationary. However, when it comes to the consensus of multi-agent systems, the corresponding results have been scattered due mainly to the difficulty in quantifying the consensus over a finite horizon. Nevertheless, the research on time-varying multi-agent systems is far from adequate especially when both nonlinear and stochastic effects become the concerns and there are still many open yet challenging problems deserving further investigation.

For deterministic nonlinear MASs, up to this date, quite a few methodologies have been exploited to cope with certain typical issues (e.g., consensus control, output regulation, tracking control, synchronization). Compared with the fruitful results on deterministic MASs, the research progress on *stochastic nonlinear* MASs has been relatively slow simply because of the coupled complexities resulting from the topology structure, the nonlinearities and the stochasticities. Among the few reported results, consensus and tracking problems have been discussed for *time-invariant* stochastic MASs subject to bounded nonlinearities, Lipschitz nonlinearities and sector-bounded nonlinearities, where the *state feedback* control has been considered. So far, despite its significant engineering insight, the *output feedback* control problem for *nonlinear time-varying stochastic* MASs has not been thoroughly investigated yet, not to mention the case where multiple performance requirements (e.g. $\mathcal{H}_\infty$ specification and mean square criterion) are also involved.

In this chapter, it is our objective to design an output-feedback controller for a class of nonlinear time-varying stochastic multi-agent systems ensuring both prespecified $\mathcal{H}_\infty$ and mean square consensus performance requirements. It should be pointed out that, different from the traditional consensus results on time-invariant systems with steady-state behaviors (e.g. asymptotic convergence of the consensus error), a new finite-horizon consensus problem is considered for the time-varying systems where we would be more interested in the transient $\mathcal{H}_2/\mathcal{H}_\infty$-type properties of the dynamic consensus against the exogenous stochastic and deterministic disturbances. In this context, three challenges are identified as follows: i) How can we propose consensus concepts capable of describing the dynamical consensus characteristics in terms of $\mathcal{H}_\infty$ and mean square specifications? ii) How can we examine the impact from the given topology, the prespecified noise attenuation level and the mean square criterion on the overall consensus performance? iii) How can we investigate the multiple performance indices (i.e. $\mathcal{H}_\infty$ and mean square indices) within a unified framework and subsequently design the corresponding multi-objective controller? We endeavor to answer the above three questions in this chapter by launching a major study on the so-called mean square $\mathcal{H}_\infty$ consensus control problem.

The rest of this chapter is organized as follows. Section 7.1 formulates the multi-objective output-feedback controller design problem for a nonlinear time-varying stochastic multi-agent system. The main results are presented in Section 7.2 where sufficient conditions for the multi-agent system reaching the desired consensus are given in terms of recursive linear matrix inequalities. Section 7.3 gives a numerical example and Section 7.4 outlines our conclusion.

7.1   Problem Formulation

In this chapter, the multi-agent system has N agents which communicate with each other according to a fixed network topology. The topology is represented by an undirected graph $\mathscr{G}$ consisting of a vertex set $\mathscr{V} = \{1, 2, \ldots, N\}$, an edge set $\Theta \in \mathscr{V} \times \mathscr{V}$ and a symmetric weighted adjacency matrix $\mathscr{H} = [h_{ij}]$ with $h_{ij} \geq 0$ where $h_{ij} > 0$ if and only if $(i, j) \in \Theta$, namely, agent i can obtain information from agent j. Self-edges (i, i) are not allowed; that is, $(i, i) \notin \Theta$ for any $i \in \mathscr{V}$. Furthermore, we denote the neighborhood of agent i by $\mathscr{N}_i = \{j \in \mathscr{V} : (j, i) \in \Theta\}$. An element of $\mathscr{N}_i$ is called a neighbor of agent i. The in-degree of agent i is defined as $\deg_{\text{in}}^i \triangleq \sum_{j \in \mathscr{N}_i} h_{ij}$.

Consider a multi-agent system with N agents where the dynamics of agent i is described by the following nonlinear discrete time-varying stochastic system defined on $k \in [0, T]$:

$$\begin{cases} x_{i,k+1} = A_k x_{i,k} + B_k u_{i,k} \\ \qquad\quad + D_k w_{i,k} + F_k f(x_{i,k}, u_{i,k}, \eta_{i,k}) \\ y_{i,k} = C_k x_{i,k} + E_k w_{i,k} \\ z_{i,k} = L_k x_{i,k} \end{cases} \tag{7.1}$$

where $x_{i,k} \in \mathbb{R}^n$, $u_{i,k} \in \mathbb{R}^p$, $y_{i,k} \in \mathbb{R}^r$ and $z_{i,k} \in \mathbb{R}^m$ are, respectively, the state, the control input, the measurement output and the controlled output of agent i. $w_{i,k} \in \mathbb{R}^\omega$ $(i = 1, 2, \ldots, N)$ represent N mutually uncorrelated zero-mean Gaussian white sequences with covariances $W_{i,k} > 0$. $\eta_{i,k} \in \mathbb{R}$ $(i = 1, 2, \ldots, N)$ are mutually uncorrelated zero-mean Gaussian white sequences that are also uncorrelated with $w_{i,k}$. A_k, B_k, C_k, D_k, E_k, F_k and L_k are time-varying matrices with compatible dimensions.

Denote $f_{i,k} \triangleq f(x_{i,k}, u_{i,k}, \eta_{i,k})$. The nonlinear stochastic functions $f_{i,k} : \mathbb{R}^n \times \mathbb{R}^p \times \mathbb{R} \mapsto \mathbb{R}^\nu$ are assumed to have the following first moments for all $x_{i,k}$ and $u_{i,k}$:

$$\mathbb{E}\{f_{i,k} | x_{i,k}, u_{i,k}\} = 0 \tag{7.2}$$

with the covariances given by

$$\mathbb{E}\{f_{i,k} f_{i,s}^{\mathrm{T}} | x_{i,k}, u_{i,k}\} = 0, \quad k \neq s$$
$$\mathbb{E}\{f_{i,k} f_{j,k}^{\mathrm{T}} | x_{i,k}, u_{i,k}\} = 0, \quad i \neq j$$
$$\mathbb{E}\{f_{i,k} f_{i,k}^{\mathrm{T}} | x_{i,k}, u_{i,k}\} \tag{7.3}$$
$$= \sum_{l=1}^{q} \Omega_{i,k}^{(l)} \left(x_{i,k}^{\mathrm{T}} \Gamma_{i,k}^{(l)} x_{i,k} + u_{i,k}^{\mathrm{T}} \Lambda_{i,k}^{(l)} u_{i,k} \right)$$

where $\Omega_{i,k}^{(l)} > 0$, $\Gamma_{i,k}^{(l)} \geq 0$ and $\Lambda_{i,k}^{(l)} \geq 0$ are known matrices of appropriate dimensions.

Remark 7.1 *The nonlinear descriptions in (7.2)–(7.3) can represent several classes of well-studied nonlinear systems such as the system with multiplicative noises (also known as state-dependent noises) and the system whose state's power depends on the sector-bound (or sign) of the nonlinear function of the state; see, e.g. [134, 239, 240].*

In this chapter, for the nonlinear time-varying stochastic multi-agent system (7.1), we adopt the control protocol of the following form:

$$u_{i,k} = K_k \sum_{j \in \mathscr{N}_i} h_{ij} \left(y_{j,k} - y_{i,k} \right) \tag{7.4}$$

where $K_k \in \mathbb{R}^{p \times r}$ $(0 \leq k \leq T)$ are the feedback gain matrices to be designed.

For the purpose of simplicity, we introduce the following notations:

$$\mathcal{H} \triangleq \begin{bmatrix} -\deg_{in}^1 & h_{12} & \cdots & h_{1N} \\ h_{21} & -\deg_{in}^2 & \cdots & h_{2N} \\ \vdots & \vdots & \ddots & \vdots \\ h_{N1} & h_{N2} & \cdots & -\deg_{in}^N \end{bmatrix},$$

$$x_k \triangleq \begin{bmatrix} x_{1,k}^{\mathrm{T}} & x_{2,k}^{\mathrm{T}} & \cdots & x_{N,k}^{\mathrm{T}} \end{bmatrix}^{\mathrm{T}},$$

$$y_k \triangleq \begin{bmatrix} y_{1,k}^{\mathrm{T}} & y_{2,k}^{\mathrm{T}} & \cdots & y_{N,k}^{\mathrm{T}} \end{bmatrix}^{\mathrm{T}},$$

$$z_k \triangleq \begin{bmatrix} z_{1,k}^{\mathrm{T}} & z_{2,k}^{\mathrm{T}} & \cdots & z_{N,k}^{\mathrm{T}} \end{bmatrix}^{\mathrm{T}},$$

$$w_k \triangleq \begin{bmatrix} w_{1,k}^{\mathrm{T}} & w_{2,k}^{\mathrm{T}} & \cdots & w_{N,k}^{\mathrm{T}} \end{bmatrix}^{\mathrm{T}}, \tag{7.5}$$

$$f_k \triangleq \begin{bmatrix} f_{1,k}^{\mathrm{T}} & f_{2,k}^{\mathrm{T}} & \cdots & f_{N,k}^{\mathrm{T}} \end{bmatrix}^{\mathrm{T}},$$

$$u_k \triangleq \begin{bmatrix} u_{1,k}^{\mathrm{T}} & u_{2,k}^{\mathrm{T}} & \cdots & u_{N,k}^{\mathrm{T}} \end{bmatrix}^{\mathrm{T}},$$

$$\mathcal{A}_k \triangleq I_N \otimes A_k + \mathcal{H} \otimes (B_k K_k C_k),$$

$$\mathcal{F}_k \triangleq (I_N \otimes F_k), \quad U \triangleq \begin{bmatrix} I_{nN} & 0 \end{bmatrix},$$

$$\mathcal{D}_k \triangleq I_N \otimes D_k + \mathcal{H} \otimes (B_k K_k E_k),$$

$$\mathcal{L}_k \triangleq (I_N \otimes L_k), \quad V \triangleq \begin{bmatrix} 0 & I_{nN} \end{bmatrix}.$$

Implementing the control law (7.4) to the multi-agent system (7.1), we have

$$x_{i,k+1} = A_k x_{i,k} + B_k K_k \sum_{j \in \mathcal{N}_i} h_{ij} (y_{j,k} - y_{i,k}) + D_k w_{i,k}$$

$$+ F_k f\left(x_{i,k}, K_k \sum_{j \in \mathcal{N}_i} h_{ij}(y_{j,k} - y_{i,k}), \eta_{i,k}\right). \tag{7.6}$$

By augmenting the system state $x_{i,k}$, we obtain the closed-loop system as follows:

$$\begin{cases} x_{k+1} = \mathcal{A}_k x_k + \mathcal{D}_k w_k + \mathcal{F}_k f_k, \\ z_k = \mathcal{L}_k x_k. \end{cases} \tag{7.7}$$

On the other hand, denote the average state of all agents by:

$$\bar{x}_k \triangleq \frac{1}{N} \sum_{i=1}^N x_{i,k} = \frac{1}{N} (1_N^{\mathrm{T}} \otimes I_n) x_k. \tag{7.8}$$

Then, taking $1_N^{\mathrm{T}} \mathcal{H} = 0$ into consideration, we obtain

$$\begin{aligned} \bar{x}_{k+1} &= \frac{1}{N} (1_N^{\mathrm{T}} \otimes I_n) x_{k+1} \\ &= \frac{1}{N} (1_N^{\mathrm{T}} \otimes I_n)\left(\mathcal{A}_k x_k + \mathcal{D}_k w_k + \mathcal{F}_k f_k\right) \\ &= A_k \bar{x}_k + \frac{1}{N} (1_N^{\mathrm{T}} \otimes D_k) w_k + \frac{1}{N} (1_N^{\mathrm{T}} \otimes F_k) f_k. \end{aligned} \tag{7.9}$$

Before proceeding further, we introduce the following definitions.

Definition 7.1 *The performance index of $\mathcal{H}_\infty$ consensus for nonlinear time-varying stochastic multi-agent system (7.1) over the finite horizon $[0,T]$ is defined by*

$$J_{[0,T]} \triangleq \frac{\mathbb{E}\left\{\sum_{i=1}^N \|z_{i,k} - \bar{z}_k\|_{[0,T]}^2\right\}}{\mathbb{E}\left\{\sum_{i=1}^N \|w_{i,k}\|_{[0,T]}^2 + \sum_{i=1}^N x_{i,0}^{\mathrm{T}} \Phi x_{i,0}\right\}} \tag{7.10}$$

where $\bar{z}_k \triangleq \frac{1}{N}\sum_{i=1}^N z_{i,k} = \frac{1}{N}(\mathbf{1}_N^{\mathrm{T}} \otimes I_m)z_k$ and $\Phi > 0$ is a given weighting matrix.

Definition 7.2 *The performance index of mean square consensus for agent i $(i = 1, 2, \ldots, N)$ of the nonlinear time-varying stochastic multi-agent system (7.1) at time instant k is defined by*

$$\mathfrak{R}_{i,k} \triangleq \mathbb{E}\{(x_{i,k} - \bar{x}_k)(x_{i,k} - \bar{x}_k)^{\mathrm{T}}\}. \tag{7.11}$$

Remark 7.2 *For time-varying multi-agent systems, we are usually interested in the transient consensus dynamics over a specified time interval. Consequently, in this chapter, we propose two consensus performance indices (in Definitions 7.1 and 7.2) to reflect the time-varying manner and characterize the transient consensus behaviors of the addressed multi-agent system. Specifically, $J_{[0,T]}$ defined in (7.10) is introduced to evaluate the noise attenuation level over the specified horizon $[0,T]$. On the other hand, the mean square performance index $\mathfrak{R}_{i,k}$ defined in (7.11), which can be intuitively understood as the "distance" from agent i to the center of the group, accounts for the deviation level of the individual agent at time instant k. It should be pointed out that, although both indices are proposed to characterize the transient performances, they actually reflect the consensus processes from different perspectives. The $\mathcal{H}_\infty$ specification is concerned with the collective consensus performance for the whole MAS over the finite horizon $[0,T]$, while the mean square criterion is put forward to depict the consensus behavior with respect to the individual agent at each time instant k.*

Assumption 7.1 *The initial values of each agent, namely, $x_{i,0}$ $(i = 1, 2, \ldots, N)$ are known and satisfying*

$$(x_{i,0} - \bar{x}_{i,0})(x_{i,0} - \bar{x}_{i,0})^{\mathrm{T}} \leq \Xi_0 \tag{7.12}$$

where $\Xi_0 > 0$ is a known positive definite matrix.

We are now in a position to state the main goal of this chapter as follows. With the given undirected communication graph $\mathscr{G}$, it is our objective to determine the output feedback gains $\{K_k\}_{0 \leq k \leq T}$ for the nonlinear time-varying stochastic MAS (7.1) such that the following requirements are met simultaneously:

(*R1*) ($\mathcal{H}_\infty$ constraint) For any nonzero w_k, the $\mathcal{H}_\infty$ consensus performance index defined in (7.10) satisfies

$$J_{[0,T]} < \gamma^2 \tag{7.13}$$

where $\gamma > 0$ is a prespecified noise attenuation level.

(*R2*) (Mean-square requirement) Under the initial condition (7.12), the mean square consensus criterion defined in (7.11) achieves

$$\mathfrak{R}_{i,k} \leq \Xi_k \ (i = 1, 2, \ldots, N; \ k \in [0, T]) \tag{7.14}$$

where $\{\Xi_k\}_{0 \leq k \leq T}$ is a prespecified sequence of positive definite matrices representing the upper bound imposed on the mean square consensus performance.

Definition 7.3 *The closed-loop nonlinear time-varying stochastic MAS (7.7) is said to reach the $(\mathcal{G}, \gamma, \Xi_k)$-dependent consensus if the design objectives (R1) and (R2) are achieved simultaneously.*

Remark 7.3 *Contrary to the traditional consensus problem for a time-invariant case where the consensus error is required to approach the desired trajectory asymptotically, the newly proposed $(\mathcal{G}, \gamma, \Xi_k)$-dependent consensus in Definition 7.3 would better reflect the transient property of the consensus process in a comprehensive way. Moreover, it should be pointed out that there exists a certain trade-off between the two essential performance criteria (γ and Ξ_k), which will be discussed in detail later in the filter design section. Such a trade-off could provide much flexibility in making a compromise between the two indices and therefore ensure a satisfactory consensus performance.*

7.2   Main Results

In this section, we first analyze the $\mathcal{H}_\infty$ and mean square consensus performances, respectively, for the addressed multi-agent system with the fixed communication graph $\mathcal{G}$. Then, by means of the recursive linear matrix inequality approach, sufficient conditions are established for the MAS (7.7) to achieve the $(\mathcal{G}, \gamma, \Xi_k)$-dependent consensus. A recursive algorithm is finally presented to compute the desired time-varying controller gain matrices $\{K_k\}_{0 \leq k \leq T}$ step by step.

First of all, we introduce the following lemmas that are useful for our further development.

Lemma 7.1 *(Schur Complement Lemma) Given constant matrices $\mathcal{S}_1, \mathcal{S}_2, \mathcal{S}_3$ where $\mathcal{S}_1 = \mathcal{S}_1^{\mathrm{T}}$ and $0 < \mathcal{S}_2 = \mathcal{S}_2^{\mathrm{T}}$, then $\mathcal{S}_1 + \mathcal{S}_3^{\mathrm{T}} \mathcal{S}_2^{-1} \mathcal{S}_3 < 0$ if and only if*

$$\begin{bmatrix} \mathcal{S}_1 & \mathcal{S}_3^{\mathrm{T}} \\ \mathcal{S}_3 & -\mathcal{S}_2 \end{bmatrix} < 0 \quad \text{or} \quad \begin{bmatrix} -\mathcal{S}_2 & \mathcal{S}_3 \\ \mathcal{S}_3^{\mathrm{T}} & \mathcal{S}_1 \end{bmatrix} < 0. \tag{7.15}$$

Lemma 7.2 *For any real-valued vectors a, b and matrix P > 0 of compatible dimensions, we have*

$$a^{\mathrm{T}} P b + b^{\mathrm{T}} P a \leq \epsilon a^{\mathrm{T}} P a + \epsilon^{-1} b^{\mathrm{T}} P b \tag{7.16}$$

where $\epsilon > 0$ is a given constant.

Denote $\tilde{x}_{i,k} \triangleq x_{i,k} - \bar{x}_k$ and $\tilde{x}_k \triangleq [\tilde{x}_{1,k}^{\mathrm{T}} \quad \tilde{x}_{2,k}^{\mathrm{T}} \quad \cdots \quad \tilde{x}_{N,k}^{\mathrm{T}}]^{\mathrm{T}}$. Subtracting (7.10) from (7.1) leads to

$$
\begin{aligned}
&\tilde{x}_{i,k+1} \\
=&\, x_{i,k+1} - \bar{x}_{k+1} \\
=&\, A_k x_{i,k} + B_k u_{i,k} + D_k w_{i,k} + F_k f(x_{i,k}) \\
&- \left(A_k \bar{x}_k + \frac{1}{N} (1_N^{\mathrm{T}} \otimes D_k) w_k + \frac{1}{N} (1_N^{\mathrm{T}} \otimes F_k) f_k \right).
\end{aligned} \tag{7.17}
$$

It can be inferred from (7.8) that

$$
\begin{aligned}
&\tilde{x}_{k+1} \\
=&\, x_{k+1} - (1_N \otimes I_n) \bar{x}_{k+1} \\
=&\, \big(I_N \otimes A_k + \mathcal{H} \otimes (B_k K_k C_k)\big) x_k \\
&+ \big(I_N \otimes D_k + \mathcal{H} \otimes (B_k K_k E_k)\big) w_k \\
&- \left((1_N \otimes A_k) \bar{x}_k + \frac{1}{N} \big((1_N 1_N^{\mathrm{T}}) \otimes D_k\big) w_k \right. \\
&\left. + \frac{1}{N} \big((1_N 1_N^{\mathrm{T}}) \otimes F_k\big) f_k \right) + (I_N \otimes F_k) f_k \\
=&\, \big(I_N \otimes A_k + \mathcal{H} \otimes (B_k K_k C_k)\big) \tilde{x}_k \\
&+ \big(\mathcal{N} \otimes D_k + \mathcal{H} \otimes (B_k K_k E_k)\big) w_k + (\mathcal{N} \otimes F_k) f_k \\
\triangleq&\, \mathcal{A}_k \tilde{x}_k + \tilde{D}_k w_k + \tilde{F}_k f_k
\end{aligned} \tag{7.18}
$$

where

$$\tilde{D}_k \triangleq \mathcal{N} \otimes D_k + \mathcal{H} \otimes (B_k K_k E_k), \quad \tilde{F}_k \triangleq \mathcal{N} \otimes F_k,$$

$$\mathcal{N} \triangleq [a_{ij}]_{N \times N} = \begin{cases} \dfrac{N-1}{N}, & i = j, \\[2mm] -\dfrac{1}{N}, & i \neq j. \end{cases}$$

Defining $\xi_k \triangleq [\, x_k^{\mathrm{T}} \quad \tilde{x}_k^{\mathrm{T}} \,]^{\mathrm{T}}$, we obtain the following augmented system:

$$\xi_{k+1} = \mathscr{A}_k \xi_k + \mathscr{D}_k w_k + \mathscr{F}_k f_k \tag{7.19}$$

where

$$\mathscr{A}_k \triangleq \begin{bmatrix} A_k & 0 \\ 0 & \mathcal{A}_k \end{bmatrix}, \quad \mathscr{D}_k \triangleq \begin{bmatrix} D_k \\ \tilde{D}_k \end{bmatrix}, \quad \mathscr{F}_k \triangleq \begin{bmatrix} F_k \\ \tilde{F}_k \end{bmatrix}.$$

$\mathcal{H}_\infty$ consensus performance

For brevity of the following notation, we define

$$\mathcal{C}_k \triangleq \mathcal{H} \otimes (K_k C_k),$$

$$\mathcal{E}_k \triangleq \mathcal{H} \otimes (K_k E_k),$$

$$W_k \triangleq \operatorname{diag}_N\{W_{i,k}\},$$

$$\Omega_k^{(l)} \triangleq \operatorname{diag}_N\{\Omega_{i,k}^{(l)}\},$$

$$\tilde{\Gamma}_{i,k}^{(l)} \triangleq \mathcal{M}_i^{\mathrm{T}} \Gamma_{i,k}^{(l)} \mathcal{M}_i,$$

$$\tilde{\Lambda}_{i,k}^{(l)} \triangleq \mathcal{Z}_i^{\mathrm{T}} \Lambda_{i,k}^{(l)} \mathcal{Z}_i,$$

$$\mathcal{I} \triangleq I_{mN} - \frac{1}{N}\big((\mathbf{1}_N \mathbf{1}_N^{\mathrm{T}} \otimes I_m)\big),$$

$$\mathcal{R}_i \triangleq \operatorname{diag}\{\underbrace{0,0,\cdots 0,}_{i-1}\ I_\nu\ \underbrace{0,\cdots,0}_{N-i}\},$$

$$\mathcal{M}_i \triangleq [\underbrace{0\ \ 0\ \ \cdots\ \ 0}_{i-1}\ \ I_n\ \underbrace{0\ \ \cdots\ \ 0}_{N-i}],$$

$$\mathcal{Z}_i \triangleq [\underbrace{0\ \ 0\ \ \cdots\ \ 0}_{i-1}\ \ I_p\ \underbrace{0\ \ \cdots\ \ 0}_{N-i}]. \tag{7.20}$$

In the following theorem, sufficient conditions are established for the closed-loop MAS (7.7) to satisfy the $\mathcal{H}_\infty$ consensus constraint by means of the recursive matrix inequality approach.

Theorem 7.1 *Consider the MAS (7.7). Let the undirected communication graph $\mathcal{G}$ and the sequence of output feedback gains $\{K_k\}_{0 \le k \le T}$ be given. For a prespecified noise attenuation level $\gamma > 0$ and a positive definite matrix $\Phi > 0$, the $\mathcal{H}_\infty$ consensus performance constraint (7.13) is achieved for all nonzero w_k if there exists a sequence of positive definite matrices $\{Q_k\}_{0 \le k \le T+1}$ with the initial condition $Q_0 \le \gamma^2 U^{\mathrm{T}}(I_N \otimes \Phi)U$ that satisfies the following recursive matrix inequality:*

$$\Psi_k \triangleq \begin{bmatrix} \Psi_{11,k} + U^{\mathrm{T}} \mathcal{L}_k^{\mathrm{T}} \mathcal{I}^2 \mathcal{L}_k U & \Psi_{12,k} \\ * & \Psi_{22,k} - \gamma^2 I_{\omega N} \end{bmatrix} < 0 \tag{7.21}$$

where

$$\mathcal{Q}_{k+1}^{(l)} \triangleq \mathscr{F}_k^{\mathrm{T}} Q_{k+1} \mathscr{F}_k \Omega_k^{(l)},$$

$$\Psi_{11,k} \triangleq \mathscr{A}_k^{\mathrm{T}} Q_{k+1} \mathscr{A}_k - Q_k$$
$$\qquad + \sum_{l=1}^{q}\sum_{i=1}^{N} \operatorname{tr}\left[\mathcal{Q}_{k+1}^{(l)} \mathcal{R}_i\right] U^{\mathrm{T}}\big(\tilde{\Gamma}_{i,k}^{(l)} + \mathcal{C}_k^{\mathrm{T}}\tilde{\Lambda}_{i,k}^{(l)}\mathcal{C}_k\big)U,$$

$$\Psi_{12,k} \triangleq \mathscr{A}_k^{\mathrm{T}} Q_{k+1} \mathscr{D}_k$$
$$\qquad + \sum_{l=1}^{q}\sum_{i=1}^{N} \operatorname{tr}\left[\mathcal{Q}_{k+1}^{(l)} \mathcal{R}_i\right] U^{\mathrm{T}}\mathcal{C}_k^{\mathrm{T}}\tilde{\Lambda}_{i,k}^{(l)}\mathcal{E}_k,$$

$$\Psi_{22,k} \triangleq \mathscr{D}_k^{\mathrm{T}} Q_{k+1} \mathscr{D}_k$$

$$+ \sum_{l=1}^{q} \sum_{i=1}^{N} \mathrm{tr}\left[\mathcal{Q}_{k+1}^{(l)} \mathcal{R}_i \right] \mathcal{E}_k^{\mathrm{T}} \tilde{\Lambda}_{i,k}^{(l)} \mathcal{E}_k. \tag{7.22}$$

Proof *By defining*

$$J_k \triangleq \xi_{k+1}^{\mathrm{T}} Q_{k+1} \xi_{k+1} - \xi_k^{\mathrm{T}} Q_k \xi_k \tag{7.23}$$

and substituting (7.19) into (7.23), we acquire

$$\begin{aligned}
\mathbb{E}\{J_k\} =& \mathbb{E}\{(\mathscr{A}_k \xi_k + \mathscr{D}_k w_k + \mathscr{F}_k f_k)^{\mathrm{T}} Q_{k+1} \\
& \times (\mathscr{A}_k \xi_k + \mathscr{D}_k w_k + \mathscr{F}_k f_k) - \xi_k^{\mathrm{T}} Q_k \xi_k \} \\
=& \mathbb{E}\{\xi_k^{\mathrm{T}} \mathscr{A}_k^{\mathrm{T}} Q_{k+1} \mathscr{A}_k \xi_k + w_k^{\mathrm{T}} \mathscr{D}_k^{\mathrm{T}} Q_{k+1} \mathscr{D}_k w_k \\
& + f_k^{\mathrm{T}} \mathscr{F}_k^{\mathrm{T}} Q_{k+1} \mathscr{F}_k f_k + \xi_k^{\mathrm{T}} \mathscr{A}_k^{\mathrm{T}} Q_{k+1} \mathscr{D}_k w_k \\
& + w_k^{\mathrm{T}} \mathscr{D}_k^{\mathrm{T}} Q_{k+1} \mathscr{A}_k \xi_k - \xi_k^{\mathrm{T}} Q_k \xi_k \}.
\end{aligned} \tag{7.24}$$

By means of the property of matrix trace, it can be obtained from (7.2) and (7.3) that

$$\begin{aligned}
& \mathbb{E}\{f_k^{\mathrm{T}} \mathscr{F}_k^{\mathrm{T}} Q_{k+1} \mathscr{F}_k f_k\} \\
=& \mathbb{E}\{\mathrm{tr}\left[\mathscr{F}_k^{\mathrm{T}} Q_{k+1} \mathscr{F}_k f_k f_k^{\mathrm{T}} \right]\} \\
=& \mathrm{tr}\left[\mathscr{F}_k^{\mathrm{T}} Q_{k+1} \mathscr{F}_k \sum_{l=1}^{q} \Omega_k^{(l)} \sum_{i=1}^{N} \mathcal{R}_i \right. \\
& \left. \times \left(x_k^{\mathrm{T}} \tilde{\Gamma}_{i,k}^{(l)} x_k + u_k^{\mathrm{T}} \tilde{\Lambda}_{i,k}^{(l)} u_k \right) \right] \\
=& x_k^{\mathrm{T}} \sum_{l=1}^{q} \sum_{i=1}^{N} \mathrm{tr}\left[\mathcal{Q}_{k+1}^{(l)} \mathcal{R}_i \right] \tilde{\Gamma}_{i,k}^{(l)} x_k \\
& + u_k^{\mathrm{T}} \sum_{l=1}^{q} \sum_{i=1}^{N} \mathrm{tr}\left[\mathcal{Q}_{k+1}^{(l)} \mathcal{R}_i \right] \tilde{\Lambda}_{i,k}^{(l)} u_k.
\end{aligned} \tag{7.25}$$

Noting (7.20) and using the Kronecker product, we can describe u_k by $u_k = \mathcal{C}_k x_k + \mathcal{E}_k w_k$ and therefore

$$\begin{aligned}
& \mathbb{E}\{f_k^{\mathrm{T}} \mathscr{F}_k^{\mathrm{T}} Q_{k+1} \mathscr{F}_k f_k\} \\
=& \xi_k^{\mathrm{T}} \sum_{l=1}^{q} \sum_{i=1}^{N} \mathrm{tr}\left[\mathcal{Q}_{k+1}^{(l)} \mathcal{R}_i \right] U^{\mathrm{T}} \tilde{\Gamma}_{i,k}^{(l)} U \xi_k \\
& + \xi_k^{\mathrm{T}} \sum_{l=1}^{q} \sum_{i=1}^{N} \mathrm{tr}\left[\mathcal{Q}_{k+1}^{(l)} \mathcal{R}_i \right] U^{\mathrm{T}} \mathcal{C}_k^{\mathrm{T}} \tilde{\Lambda}_{i,k}^{(l)} \mathcal{C}_k U \xi_k \\
& + \xi_k^{\mathrm{T}} \sum_{l=1}^{q} \sum_{i=1}^{N} \mathrm{tr}\left[\mathcal{Q}_{k+1}^{(l)} \mathcal{R}_i \right] U^{\mathrm{T}} \mathcal{C}_k^{\mathrm{T}} \tilde{\Lambda}_{i,k}^{(l)} \mathcal{E}_k w_k
\end{aligned}$$

$$+ w_k^{\mathrm{T}} \sum_{l=1}^{q} \sum_{i=1}^{N} \mathrm{tr}\left[\mathcal{Q}_{k+1}^{(l)} \mathcal{R}_i\right] \mathcal{E}_k^{\mathrm{T}} \tilde{\Lambda}_{i,k}^{(l)} \mathcal{C}_k U \xi_k$$

$$+ w_k^{\mathrm{T}} \sum_{l=1}^{q} \sum_{i=1}^{N} \mathrm{tr}\left[\mathcal{Q}_{k+1}^{(l)} \mathcal{R}_i\right] \mathcal{E}_k^{\mathrm{T}} \tilde{\Lambda}_{i,k}^{(l)} \mathcal{E}_k w_k. \tag{7.26}$$

Consequently,

$$\begin{aligned}
\mathbb{E}\{J_k\} =& \mathbb{E}\{(\mathscr{A}_k \xi_k + \mathscr{D}_k w_k + \mathscr{F}_k f_k)^{\mathrm{T}} Q_{k+1} \\
& \times (\mathscr{A}_k \xi_k + \mathscr{D}_k w_k + \mathscr{F}_k f_k) - \xi_k^{\mathrm{T}} Q_k \xi_k\} \\
=& \mathbb{E}\Bigg\{ \xi_k^{\mathrm{T}} \Big(\mathscr{A}_k^{\mathrm{T}} Q_{k+1} \mathscr{A}_k - Q_k \\
& + \sum_{l=1}^{q} \sum_{i=1}^{N} \mathrm{tr}\left[\mathcal{Q}_{k+1}^{(l)} \mathcal{R}_i\right] U^{\mathrm{T}} \tilde{\Gamma}_{i,k}^{(l)} U \\
& + \sum_{l=1}^{q} \sum_{i=1}^{N} \mathrm{tr}\left[\mathcal{Q}_{k+1}^{(l)} \mathcal{R}_i\right] U^{\mathrm{T}} \mathcal{C}_k^{\mathrm{T}} \tilde{\Lambda}_{i,k}^{(l)} \mathcal{C}_k U \Big) \xi_k \\
& + w_k^{\mathrm{T}} \Big(\mathscr{D}_k^{\mathrm{T}} Q_{k+1} \mathscr{D}_k \\
& + \sum_{l=1}^{q} \sum_{i=1}^{N} \mathrm{tr}\left[\mathcal{Q}_{k+1}^{(l)} \mathcal{R}_i\right] \mathcal{E}_k^{\mathrm{T}} \tilde{\Lambda}_{i,k}^{(l)} \mathcal{E}_k \Big) w_k \\
& + \xi_k^{\mathrm{T}} \Big(\mathscr{A}_k^{\mathrm{T}} Q_{k+1} \mathscr{D}_k \\
& + \sum_{l=1}^{q} \sum_{i=1}^{N} \mathrm{tr}\left[\mathcal{Q}_{k+1}^{(l)} \mathcal{R}_i\right] U^{\mathrm{T}} \mathcal{C}_k^{\mathrm{T}} \tilde{\Lambda}_{i,k}^{(l)} \mathcal{E}_k \Big) w_k \\
& + w_k^{\mathrm{T}} \Big(\mathscr{D}_k^{\mathrm{T}} Q_{k+1} \mathscr{A}_k \\
& + \sum_{l=1}^{q} \sum_{i=1}^{N} \mathrm{tr}\left[\mathcal{Q}_{k+1}^{(l)} \mathcal{R}_i\right] \mathcal{E}_k^{\mathrm{T}} \tilde{\Lambda}_{i,k}^{(l)} \mathcal{C}_k U \Big) \xi_k \Bigg\} \\
=& \mathbb{E}\left\{ \begin{bmatrix} \xi_k \\ w_k \end{bmatrix}^{\mathrm{T}} \begin{bmatrix} \Psi_{11,k} & \Psi_{12,k} \\ * & \Psi_{22,k} \end{bmatrix} \begin{bmatrix} \xi_k \\ w_k \end{bmatrix} \right\} \tag{7.27}
\end{aligned}$$

where $\Psi_{11,k}$, $\Psi_{12,k}$ and $\Psi_{22,k}$ are defined in (7.22).

On the other hand, it is easy to see that the index of $\mathcal{H}_\infty$ consensus performance defined in (7.10) can be equivalently expressed by

$$\begin{aligned}
J_{[0,T]} =& \frac{\mathbb{E}\left\{\|z_k - (\mathbf{1}_N \otimes I_m)\bar{z}_k\|_{[0,T]}^2\right\}}{\mathbb{E}\left\{\|w_k\|_{[0,T]}^2 + x_0^{\mathrm{T}}(I_N \otimes \Phi)x_0\right\}} \\
=& \frac{\mathbb{E}\left\{\|z_k^*\|_{[0,T]}^2\right\}}{\mathbb{E}\left\{\|w_k\|_{[0,T]}^2 + x_0^{\mathrm{T}}(I_N \otimes \Phi)x_0\right\}} \tag{7.28}
\end{aligned}$$

where $z_k^* \triangleq \mathcal{I} z_k$ with $\mathcal{I}$ being defined in (7.20).

Subsequently, adding the zero term $\|z_k^*\|^2 - \gamma^2 \|w_k\|^2 - (\|z_k^*\|^2 - \gamma^2 \|w_k\|^2)$ to $\mathbb{E}\{J_k\}$ results in

$$
\begin{aligned}
\mathbb{E}\{J_k\} =& \xi_{k+1}^{\mathrm{T}} Q_{k+1} \xi_{k+1} - \xi_k^{\mathrm{T}} Q_k \xi_k \\
=& \mathbb{E}\left\{ \begin{bmatrix} \xi_k^{\mathrm{T}} & w_k^{\mathrm{T}} \end{bmatrix} \Psi_k \begin{bmatrix} \xi_k \\ w_k \end{bmatrix} \right. \\
& \left. - (\|z_k^*\|^2 - \gamma^2 \|w_k\|^2) \right\}
\end{aligned}
\tag{7.29}
$$

where Ψ_k is defined in (7.22).

Summing (7.29) on both sides from 0 to T with respect to k yields

$$
\begin{aligned}
\sum_{k=0}^{T} \mathbb{E}\{J_k\} =& \mathbb{E}\{\xi_{T+1}^{\mathrm{T}} Q_{T+1} \xi_{T+1} - \xi_0^{\mathrm{T}} Q_0 \xi_0\} \\
=& \mathbb{E}\left\{ \sum_{k=0}^{T} \begin{bmatrix} \xi_k^{\mathrm{T}} & w_k^{\mathrm{T}} \end{bmatrix} \Psi_k \begin{bmatrix} \xi_k \\ w_k \end{bmatrix} \right\} \\
& - \mathbb{E}\left\{ \sum_{k=0}^{T} (\|z_k^*\|^2 - \gamma^2 \|w_k\|^2) \right\}
\end{aligned}
\tag{7.30}
$$

and therefore

$$
\begin{aligned}
& \mathbb{E}\left\{ \sum_{k=0}^{T} (\|z_k^*\|^2 - \gamma^2 \|w_k\|^2) \right\} - \gamma^2 x_0^{\mathrm{T}} (I_N \otimes \Phi) x_0 \\
=& \mathbb{E}\left\{ \sum_{k=0}^{T} \begin{bmatrix} \xi_k^{\mathrm{T}} & w_k^{\mathrm{T}} \end{bmatrix} \Psi_k \begin{bmatrix} \xi_k \\ w_k \end{bmatrix} \right\} \\
& - \mathbb{E}\{\xi_{T+1}^{\mathrm{T}} Q_{T+1} \xi_{T+1} \\
& - \xi_0^{\mathrm{T}} (Q_0 - \gamma^2 U^{\mathrm{T}} (I_N \otimes \Phi) U) \xi_0\}.
\end{aligned}
\tag{7.31}
$$

Since $\Psi_k < 0$ and $Q_{T+1} > 0$, it follows from the initial condition $Q_0 \leq \gamma^2 U^{\mathrm{T}} (I_N \otimes \Phi) U$ that

$$
\mathbb{E}\left\{ \|z_k^*\|_{[0,T]}^2 - \gamma^2 \|w_k\|_{[0,T]}^2 \right\} < \gamma^2 x_0^{\mathrm{T}} (I_N \otimes \Phi) x_0,
\tag{7.32}
$$

which demonstrates that $J_{[0,T]} < \gamma^2$ is satisfied. The proof is now complete.

Mean-square consensus performance

In this subsection, we are going to examine the mean square consensus performance for the closed-loop multi-agent system (7.7).

It is obvious that the mean square consensus performance defined in (7.11) can be equivalently expressed as follows:

$$\begin{aligned}
\Re_{i,k} &= \mathbb{E}\{\tilde{x}_{i,k}\tilde{x}_{i,k}^{\mathrm{T}}\} \\
&= \mathbb{E}\{\mathcal{M}_i\tilde{x}_k\tilde{x}_k^{\mathrm{T}}\mathcal{M}_i^{\mathrm{T}}\} \\
&= \mathbb{E}\{\mathcal{M}_iV\xi_k\xi_k^{\mathrm{T}}V^{\mathrm{T}}\mathcal{M}_i^{\mathrm{T}}\}.
\end{aligned} \tag{7.33}$$

By defining $\Sigma_k \triangleq \mathbb{E}\{\xi_k\xi_k^{\mathrm{T}}\}$, we can further acquire

$$\Re_{i,k} = \mathcal{M}_iV\Sigma_kV^{\mathrm{T}}\mathcal{M}_i^{\mathrm{T}}. \tag{7.34}$$

In terms of recursive matrix inequality, the following theorem presents sufficient conditions for the existence of the upper-bound on Σ_k, which plays an essential role in mean square consensus analysis and the subsequent controller design.

Theorem 7.2 *Consider the MAS (7.7). Let the undirected communication graph $\mathscr{G}$ and the sequence of output feedback gains $\{K_k\}_{0\leq k\leq T}$ be given. If there exists a sequence of positive definite matrices $\{P_k\}_{0\leq k\leq T+1}$ satisfying the following recursive matrix inequality:*

$$\begin{aligned}
P_{k+1} \geq{}& \mathscr{A}_kP_k\mathscr{A}_k^{\mathrm{T}} + \mathscr{D}_kW_k\mathscr{D}_k^{\mathrm{T}} \\
&+ \sum_{l=1}^{q}\sum_{i=1}^{N}\mathscr{F}_k\Omega_k^{(l)}\mathcal{R}_i\Big(\mathrm{tr}\big[\tilde{\Gamma}_{i,k}^{(l)}UP_kU^{\mathrm{T}}\big] \\
&+ \mathrm{tr}\big[\tilde{\Lambda}_{i,k}^{(l)}\big(C_kUP_kU^{\mathrm{T}}C_k^{\mathrm{T}} + \mathcal{E}_kW_k\mathcal{E}_k^{\mathrm{T}}\big)\big]\Big)\mathscr{F}_k^{\mathrm{T}}
\end{aligned} \tag{7.35}$$

with the initial condition $P_0 = \Sigma_0$, then $P_k \geq \Sigma_k$ holds for all $0 \leq k \leq T$.

Proof *First, let us derive the Lyapunov-like equation that governs the evolution of Σ_k over the finite horizon $[0, T]$. By taking the augmented system (7.19) into consideration, we can calculate Σ_{k+1} as follows:*

$$\begin{aligned}
\Sigma_{k+1} ={}& \mathbb{E}\{\xi_{k+1}\xi_{k+1}^{\mathrm{T}}\} \\
={}& \mathbb{E}\{(\mathscr{A}_k\xi_k + \mathscr{D}_kw_k + \mathscr{F}_kf_k) \\
&\times (\mathscr{A}_k\xi_k + \mathscr{D}_kw_k + \mathscr{F}_kf_k)^{\mathrm{T}}\} \\
={}& \mathscr{A}_k\mathbb{E}\{\xi_k\xi_k^{\mathrm{T}}\}\mathscr{A}_k^{\mathrm{T}} + \mathscr{D}_k\mathbb{E}\{w_kw_k^{\mathrm{T}}\}\mathscr{D}_k^{\mathrm{T}} \\
&+ \mathscr{F}_k\mathbb{E}\{f_kf_k^{\mathrm{T}}\}\mathscr{F}_k^{\mathrm{T}}.
\end{aligned} \tag{7.36}$$

Noticing (7.3), we have

$$\mathbb{E}\{f_k f_k^{\mathrm{T}}\}$$

$$=\mathbb{E}\left\{\sum_{l=1}^{q}\Omega_k^{(l)}\sum_{i=1}^{N}\mathcal{R}_i\big((\mathcal{M}_i x_k)^{\mathrm{T}}\Gamma_{i,k}^{(l)}\mathcal{M}_i x_k + (\mathcal{Z}_i u_k)^{\mathrm{T}}\Lambda_{i,k}^{(l)}\mathcal{Z}_i u_k\big)\right\}$$

$$=\mathbb{E}\left\{\sum_{l=1}^{q}\sum_{i=1}^{N}\Omega_k^{(l)}\mathcal{R}_i\Big(\mathrm{tr}\big[\tilde{\Gamma}_{i,k}^{(l)}U\xi_k\xi_k^{\mathrm{T}}U^{\mathrm{T}}\big]\right.$$

$$\left.+\,\mathrm{tr}\big[\tilde{\Lambda}_{i,k}^{(l)}\mathcal{C}_k U\xi_k\xi_k^{\mathrm{T}}U^{\mathrm{T}}\mathcal{C}_k^{\mathrm{T}}\big] + \mathrm{tr}\big[\tilde{\Lambda}_{i,k}^{(l)}\mathcal{E}_k W_k\mathcal{E}_k^{\mathrm{T}}\big]\Big)\right\}$$

$$=\sum_{l=1}^{q}\sum_{i=1}^{N}\Omega_k^{(l)}\mathcal{R}_i\Big(\mathrm{tr}\big[\tilde{\Gamma}_{i,k}^{(l)}U\Sigma_k U^{\mathrm{T}}\big]$$

$$+\,\mathrm{tr}\big[\tilde{\Lambda}_{i,k}^{(l)}\mathcal{C}_k U\Sigma_k U^{\mathrm{T}}\mathcal{C}_k^{\mathrm{T}}\big] + \mathrm{tr}\big[\tilde{\Lambda}_{i,k}^{(l)}\mathcal{E}_k W_k\mathcal{E}_k^{\mathrm{T}}\big]\Big). \tag{7.37}$$

Accordingly,

$$\Sigma_{k+1} =\mathscr{A}_k\Sigma_k\mathscr{A}_k^{\mathrm{T}} + \mathscr{D}_k W_k\mathscr{D}_k^{\mathrm{T}}$$

$$+\sum_{l=1}^{q}\sum_{i=1}^{N}\mathscr{F}_k\Omega_k^{(l)}\mathcal{R}_i\Big(\mathrm{tr}\big[\tilde{\Gamma}_{i,k}^{(l)}U\Sigma_k U^{\mathrm{T}}\big]$$

$$+\,\mathrm{tr}\big[\tilde{\Lambda}_{i,k}^{(l)}\big(\mathcal{C}_k U\Sigma_k U^{\mathrm{T}}\mathcal{C}_k^{\mathrm{T}} + \mathcal{E}_k W_k\mathcal{E}_k^{\mathrm{T}}\big)\big]\Big)\mathscr{F}_k^{\mathrm{T}}. \tag{7.38}$$

The rest of the proof is performed by induction. When $k = 0$, it is obvious that $P_0 \geq \Sigma_0$ holds. Supposing that $P_k \geq \Sigma_k$ for $k > 0$; then according to the principle of induction, it remains to demonstrate that $P_{k+1} \geq \Sigma_{k+1}$ is true with the condition given in the theorem. In fact, considering the property of matrix trace, we know from the matrix inequality (7.35) and $P_k \geq \Sigma_k$ that

$$P_{k+1} \geq\mathscr{A}_k P_k\mathscr{A}_k^{\mathrm{T}} + \mathscr{D}_k W_k\mathscr{D}_k^{\mathrm{T}}$$

$$+\sum_{l=1}^{q}\sum_{i=1}^{N}\mathscr{F}_k\Omega_k^{(l)}\mathcal{R}_i\Big(\mathrm{tr}\big[\tilde{\Gamma}_{i,k}^{(l)}U P_k U^{\mathrm{T}}\big]$$

$$+\,\mathrm{tr}\big[\tilde{\Lambda}_{i,k}^{(l)}\big(\mathcal{C}_k U P_k U^{\mathrm{T}}\mathcal{C}_k^{\mathrm{T}} + \mathcal{E}_k W_k\mathcal{E}_k^{\mathrm{T}}\big)\big]\Big)\mathscr{F}_k^{\mathrm{T}}$$

$$\geq\mathscr{A}_k\Sigma_k\mathscr{A}_k^{\mathrm{T}} + \mathscr{D}_k W_k\mathscr{D}_k^{\mathrm{T}}$$

$$+\sum_{l=1}^{q}\sum_{i=1}^{N}\mathscr{F}_k\Omega_k^{(l)}\mathcal{R}_i\Big(\mathrm{tr}\big[\tilde{\Gamma}_{i,k}^{(l)}U\Sigma_k U^{\mathrm{T}}\big]$$

$$+\,\mathrm{tr}\big[\tilde{\Lambda}_{i,k}^{(l)}\big(\mathcal{C}_k U\Sigma_k U^{\mathrm{T}}\mathcal{C}_k^{\mathrm{T}} + \mathcal{E}_k W_k\mathcal{E}_k^{\mathrm{T}}\big)\big]\Big)\mathscr{F}_k^{\mathrm{T}}$$

$$=\Sigma_{k+1}, \tag{7.39}$$

which indicates that the induction is accomplished and the proof of the theorem is now complete.

$(\mathscr{G}, \gamma, \Xi_k)$-dependent consensus

Based on the previous analysis, we are now ready to derive the sufficient conditions for the addressed closed-loop MAS (7.7) to reach the desired $(\mathscr{G}, \gamma, \Xi_k)$-dependent consensus with respect to the fixed γ and $\{\Xi_k\}_{0 \leq k \leq T}$. First of all, without loss of generality, we assume that the matrices $\Omega_k^{(l)}$, $\Gamma_{i,k}^{(l)}$ and $\Lambda_{i,k}^{(l)}$ can be decomposed by

$$\Omega_k^{(l)} = \mathrm{diag}_N\{\Omega_{i,k}^{(l)}\} = \mathrm{col}_N\{\eta_{i,k}^{(l)}(\eta_{i,k}^{(l)})^{\mathrm{T}}\} \triangleq \eta_k^{(l)}(\eta_k^{(l)})^{\mathrm{T}},$$

$$\Gamma_{i,k}^{(l)} = \vartheta_{i,k}^{(l)}(\vartheta_{i,k}^{(l)})^{\mathrm{T}}, \quad \Lambda_{i,k}^{(l)} = \lambda_{i,k}^{(l)}(\lambda_{i,k}^{(l)})^{\mathrm{T}} \tag{7.40}$$

where $\eta_k^{(l)} \in \mathbb{R}^{\nu N}$, $\vartheta_{i,k}^{(l)} \in \mathbb{R}^n$ and $\lambda_{i,k}^{(l)} \in \mathbb{R}^p$ are column vectors. Moreover, we denote

$$\mathcal{M} \triangleq \begin{bmatrix} \mathcal{M}_1^{\mathrm{T}} & \mathcal{M}_2^{\mathrm{T}} & \cdots & \mathcal{M}_N^{\mathrm{T}} \end{bmatrix},$$

$$\mathcal{Z} \triangleq \begin{bmatrix} \mathcal{Z}_1^{\mathrm{T}} & \mathcal{Z}_2^{\mathrm{T}} & \cdots & \mathcal{Z}_N^{\mathrm{T}} \end{bmatrix},$$

$$\mathcal{U}_k \triangleq U^{\mathrm{T}}\mathcal{M} \begin{bmatrix} \mathrm{diag}_N\{\vartheta_{i,k}^{(1)}\} & \cdots & \mathrm{diag}_N\{\vartheta_{i,k}^{(q)}\} \end{bmatrix},$$

$$\mathcal{C}_k \triangleq U^{\mathrm{T}}\mathcal{C}_k^{\mathrm{T}}\mathcal{Z} \begin{bmatrix} \mathrm{diag}_N\{\lambda_{i,k}^{(1)}\} & \cdots & \mathrm{diag}_N\{\lambda_{i,k}^{(q)}\} \end{bmatrix},$$

$$\mathcal{E}_k \triangleq \mathcal{E}_k^{\mathrm{T}}\mathcal{Z} \begin{bmatrix} \mathrm{diag}_N\{\lambda_{i,k}^{(1)}\} & \cdots & \mathrm{diag}_N\{\lambda_{i,k}^{(q)}\} \end{bmatrix},$$

$$\Delta_k \triangleq \mathrm{diag}\{\tilde{\alpha}^{(1)}, \tilde{\alpha}^{(2)}, \ldots, \tilde{\alpha}^{(q)}\},$$

$$\tilde{\alpha}^{(l)} \triangleq \mathrm{diag}\{\alpha_{1,k}^{(l)}, \alpha_{2,k}^{(l)}, \ldots, \alpha_{N,k}^{(l)}\}, \quad l = 1, 2, \ldots, q. \tag{7.41}$$

Theorem 7.3 *Let the triple $(\mathscr{G}, \gamma, \{\Xi_k\}_{0 \leq k \leq T})$, the positive-definite matrix Φ and the sequence of output feedback gains $\{K_k\}_{0 \leq k \leq T}$ be given. Let $\epsilon > 0$ be any positive constant. The closed-loop nonlinear time-varying stochastic multi-agent system (7.7) achieves the $(\mathscr{G}, \gamma, \Xi_k)$-dependent consensus if there exist two sequences of positive scalars $\{\alpha_{i,k}^{(l)}\}_{0 \leq k \leq T}$ and $\{\beta_{i,k}^{(l)}\}_{0 \leq k \leq T}$ $(i = 1, 2, \ldots, N; l = 1, 2, \ldots, q)$, two sequences of positive definite matrices $\{Q_k\}_{0 \leq k \leq T+1}$ and $\{P_k\}_{0 \leq k \leq T+1}$, with the initial values $Q_0 \leq \gamma^2 U^{\mathrm{T}}(I_N \otimes \Phi)U$ and $P_0 = \Sigma_0$, satisfying the following recursive matrix inequalities:*

$$\begin{bmatrix} -\alpha_{i,k}^{(l)} & \alpha_{i,k}^{(l)}(\eta_k^{(l)})^{\mathrm{T}}\mathcal{R}_i\mathscr{F}_k^{\mathrm{T}} \\ * & -Q_{k+1}^{-1} \end{bmatrix} < 0, \tag{7.42}$$

$$\begin{bmatrix} -Q_k & 0 & \mathscr{A}_k^{\mathrm{T}} & \mathcal{S}_{1,k} \\ * & -\gamma^2 I_{\omega N} & \mathscr{D}_k^{\mathrm{T}} & \mathcal{S}_{2,k} \\ * & * & -Q_{k+1}^{-1} & 0 \\ * & * & * & -\tilde{\Delta}_k \end{bmatrix} < 0 \tag{7.43}$$

$$\begin{bmatrix} -\beta_{i,k}^{(l)} & (\lambda_{i,k}^{(l)})^{\mathrm{T}}\mathcal{Z}_i\mathcal{C}_kU & (\lambda_{i,k}^{(l)})^{\mathrm{T}}\mathcal{Z}_i\mathcal{E}_k \\ * & -P_k^{-1} & 0 \\ * & * & -W_k^{-1} \end{bmatrix} \leq 0, \qquad (7.44)$$

$$\begin{bmatrix} -P_{k+1}+\mathcal{S}_{3,k} & \mathscr{A}_k & \mathscr{D}_k \\ * & -P_k^{-1} & 0 \\ * & * & -W_k^{-1} \end{bmatrix} \leq 0, \qquad (7.45)$$

$$\begin{bmatrix} -\Xi_{k+1} & \mathcal{M}_iVP_{k+1} \\ * & -P_{k+1} \end{bmatrix} \leq 0 \qquad (7.46)$$

where

$$\mathcal{S}_{1,k} \triangleq \begin{bmatrix} \mathscr{U}_k & \sqrt{1+\epsilon}\mathscr{C}_k & 0 & U^{\mathrm{T}}\mathcal{L}_k^{\mathrm{T}}\mathcal{I} \end{bmatrix},$$
$$\mathcal{S}_{2,k} \triangleq \begin{bmatrix} 0 & 0 & \sqrt{1+\epsilon^{-1}}\mathscr{E}_k & 0 \end{bmatrix},$$
$$\mathcal{S}_{3,k} \triangleq \sum_{l=1}^{q}\sum_{i=1}^{N}\mathscr{F}_k\Omega_k^{(l)}\mathcal{R}_i\Big(\mathrm{tr}\big[\tilde{\Gamma}_{i,k}^{(l)}UP_kU^{\mathrm{T}}\big]+\beta_{i,k}^{(l)}\Big)\mathscr{F}_k^{\mathrm{T}},$$
$$\tilde{\Delta}_k \triangleq \mathrm{diag}\{\Delta_k,\Delta_k,\Delta_k,I_{mN}\}.$$

Proof *We first prove that inequalities (7.42) and (7.43) imply the inequality (7.21), and inequalities (7.44) and (7.45) indicate the inequality (7.35).*

By Schur Complement Lemma (Lemma 7.1), the inequality (7.42) holds if and only if

$$-(\alpha_{i,k}^{(l)})^{-1} + (\eta_k^{(l)})^{\mathrm{T}}\mathcal{R}_i\mathscr{F}_k^{\mathrm{T}}Q_{k+1}\mathscr{F}_k\mathcal{R}_i\eta_k^{(l)} < 0 \qquad (7.47)$$

which, by the property of matrix trace, is equivalent to

$$\mathrm{tr}\big[\mathscr{F}_k^{\mathrm{T}}Q_{k+1}\mathscr{F}_k\mathcal{R}_i\eta_k^{(l)}(\eta_k^{(l)})^{\mathrm{T}}\mathcal{R}_i\big]$$
$$=\mathrm{tr}\big[\mathscr{F}_k^{\mathrm{T}}Q_{k+1}\mathscr{F}_k\Omega_k^{(l)}\mathcal{R}_i\big]$$
$$<(\alpha_{i,k}^{(l)})^{-1}. \qquad (7.48)$$

Similarly, it follows from Lemma 7.1 that the inequality (7.43) is true if and only if

$$\begin{bmatrix} \bar{\Psi}_{11,k} & \mathscr{A}_k^{\mathrm{T}}Q_{k+1}\mathscr{D}_k \\ \mathscr{D}_k^{\mathrm{T}}Q_{k+1}\mathscr{A}_k & \bar{\Psi}_{22,k} \end{bmatrix} < 0 \qquad (7.49)$$

where

$$\bar{\Psi}_{11,k} \triangleq \mathscr{A}_k^{\mathrm{T}}Q_{k+1}\mathscr{A}_k - Q_k + U^{\mathrm{T}}\mathcal{L}_k^{\mathrm{T}}\mathcal{I}^2\mathcal{L}_kU$$
$$+ \sum_{l=1}^{q}\sum_{i=1}^{N}U^{\mathrm{T}}\big(\tilde{\Gamma}_{i,k}^{(l)} + (1+\epsilon)\mathcal{C}_k^{\mathrm{T}}\tilde{\Lambda}_{i,k}^{(l)}\mathcal{C}_k\big)U(\alpha_{i,k}^{(l)})^{-1},$$
$$\bar{\Psi}_{22,k} \triangleq \mathscr{D}_k^{\mathrm{T}}Q_{k+1}\mathscr{D}_k - \gamma^2 I_{\omega N}$$
$$+ \sum_{l=1}^{q}\sum_{i=1}^{N}(1+\epsilon^{-1})\mathcal{E}_k^{\mathrm{T}}\tilde{\Lambda}_{i,k}^{(l)}\mathcal{E}_k(\alpha_{i,k}^{(l)})^{-1}.$$

It can be inferred from Lemma 7.2 that, for any $\epsilon > 0$, the following derivation is true:

$$\xi_k^{\mathrm{T}} \sum_{l=1}^{q} \sum_{i=1}^{N} \mathrm{tr}\left[\mathcal{Q}_{k+1}^{(l)} \mathcal{R}_i\right] U^{\mathrm{T}} \mathcal{C}_k^{\mathrm{T}} \tilde{\Lambda}_{i,k}^{(l)} \mathcal{E}_k w_k$$

$$+ w_k^{\mathrm{T}} \sum_{l=1}^{q} \sum_{i=1}^{N} \mathrm{tr}\left[\mathcal{Q}_{k+1}^{(l)} \mathcal{R}_i\right] \mathcal{E}_k^{\mathrm{T}} \tilde{\Lambda}_{i,k}^{(l)} \mathcal{C}_k U \xi_k$$

$$\leq \epsilon \xi_k^{\mathrm{T}} \sum_{l=1}^{q} \sum_{i=1}^{N} \mathrm{tr}\left[\mathcal{Q}_{k+1}^{(l)} \mathcal{R}_i\right] U^{\mathrm{T}} \mathcal{C}_k^{\mathrm{T}} \tilde{\Lambda}_{i,k}^{(l)} \mathcal{C}_k U \xi_k$$

$$+ \epsilon^{-1} w_k^{\mathrm{T}} \sum_{l=1}^{q} \sum_{i=1}^{N} \mathrm{tr}\left[\mathcal{Q}_{k+1}^{(l)} \mathcal{R}_i\right] \mathcal{E}_k^{\mathrm{T}} \tilde{\Lambda}_{i,k}^{(l)} \mathcal{E}_k w_k. \qquad (7.50)$$

Consequently, taking both (7.49) and (7.50) into consideration, we know that the inequality (7.21) can be implied by inequalities (7.42) and (7.43), which, according to Theorem 7.1, indicates that the required $\mathcal{H}_\infty$ consensus performance (Requirement (R1)) is guaranteed.

By the same token, applying the Schur Complement Lemma to the inequality (7.44) leads to

$$-\beta_{i,k}^{(l)} + (\lambda_{i,k}^{(l)})^{\mathrm{T}} \mathcal{Z}_i \mathcal{C}_k U P_k U^{\mathrm{T}} \mathcal{C}_k^{\mathrm{T}} \mathcal{Z}_i^{\mathrm{T}} \lambda_{i,k}^{(l)}$$
$$+ (\lambda_{i,k}^{(l)})^{\mathrm{T}} \mathcal{Z}_i \mathcal{E}_k W_k \mathcal{E}_k^{\mathrm{T}} \mathcal{Z}_i^{\mathrm{T}} \lambda_{i,k}^{(l)} \leq 0 \qquad (7.51)$$

or, equivalently,

$$\mathrm{tr}\left[(\mathcal{C}_k U P_k U^{\mathrm{T}} \mathcal{C}_k^{\mathrm{T}} + \mathcal{E}_k W_k \mathcal{E}_k^{\mathrm{T}}) \mathcal{Z}_i^{\mathrm{T}} \lambda_{i,k}^{(l)} (\lambda_{i,k}^{(l)})^{\mathrm{T}} \mathcal{Z}_i\right]$$
$$= \mathrm{tr}\left[\tilde{\Lambda}_{i,k}^{(l)} (\mathcal{C}_k U P_k U^{\mathrm{T}} \mathcal{C}_k^{\mathrm{T}} + \mathcal{E}_k W_k \mathcal{E}_k^{\mathrm{T}})\right]$$
$$\leq \beta_{i,k}^{(l)}. \qquad (7.52)$$

Subsequently, using Lemma 7.1 again with the inequality (7.45) and noting (7.52), we acquire that the inequality (7.35) is implied by (7.44) and (7.45). Thus, it follows from Theorem 7.2 that $\Sigma_k \leq P_k$ holds for all $k \in [0, T]$. Accordingly, by Schur Complement Lemma, it can be derived from (7.34) and (7.46) that

$$\mathfrak{R}_{i,k+1} = \mathcal{M}_i V \Sigma_{k+1} V^{\mathrm{T}} \mathcal{M}_i^{\mathrm{T}}$$
$$\leq \mathcal{M}_i V P_{k+1} V^{\mathrm{T}} \mathcal{M}_i^{\mathrm{T}}$$
$$\leq \Xi_{k+1}, \qquad (7.53)$$

which means that the required mean square consensus performance (Requirement (R2)) is also met. Therefore, the closed-loop multi-agent system (7.7) achieves the desired $(\mathcal{G}, \gamma, \Xi_k)$-dependent consensus. The proof is now complete.

Controller design

In this subsection, we are going to present the solution to the mean square $\mathcal{H}_\infty$ consensus control design problem by resorting to the recursive linear matrix inequality approach.

Theorem 7.4 *Let the triple* $(\mathscr{G}, \gamma, \{\Xi_k\}_{0 \leq k \leq T})$ *and the positive-definite matrix* Φ *be given. Let* $\epsilon > 0$ *be any positive constant. The closed-loop nonlinear time-varying stochastic multi-agent system (7.7) achieves* $(\mathscr{G}, \gamma, \Xi_k)$-*dependent consensus if there exist two sequences of positive scalars* $\{\alpha_{i,k}^{(l)}\}_{0 \leq k \leq T}$ *and* $\{\beta_{i,k}^{(l)}\}_{0 \leq k \leq T}$ $(l = 1, 2, \ldots, q; i = 1, 2, \ldots, N)$, *a sequence of real-valued matrices* $\{K_k\}_{0 \leq k \leq T}$, *sequences of positive definite matrices* $\{\mathscr{Q}_k\}_{1 \leq k \leq T+1}$ *and* $\{\mathscr{P}_k\}_{1 \leq k \leq T+1}$ *satisfying the following recursive linear matrix inequalities:*

$$\begin{bmatrix} -\alpha_{i,k}^{(l)} & \alpha_{i,k}^{(l)} \mathrm{col}_N\{\eta_{i,k}^{(l)}\}^{\mathrm{T}} \mathcal{R}_i \mathscr{F}_k^{\mathrm{T}} \\ * & -\mathscr{Q}_{k+1} \end{bmatrix} < 0, \tag{7.54}$$

$$\begin{bmatrix} -X_k & 0 & \mathscr{A}_k^{\mathrm{T}} & \mathcal{S}_{1,k} \\ * & -\gamma^2 I_{\omega N} & \mathscr{D}_k^{\mathrm{T}} & \mathcal{S}_{2,k} \\ * & * & -\mathscr{Q}_{k+1} & 0 \\ * & * & * & -\Delta_k \end{bmatrix} < 0, \tag{7.55}$$

$$\begin{bmatrix} -\beta_{i,k}^{(l)} & (\lambda_{i,k}^{(l)})^{\mathrm{T}} \mathcal{Z}_i \mathcal{C}_k U & (\lambda_{i,k}^{(l)})^{\mathrm{T}} \mathcal{Z}_i \mathcal{E}_k \\ * & -Y_k & 0 \\ * & * & -W_k^{-1} \end{bmatrix} \leq 0, \tag{7.56}$$

$$\begin{bmatrix} -\mathscr{P}_{k+1} + \tilde{\mathcal{S}}_{3,k} & \mathscr{A}_k & \mathscr{D}_k \\ * & -Y_k & 0 \\ * & * & -W_k^{-1} \end{bmatrix} \leq 0, \tag{7.57}$$

$$\begin{bmatrix} -\Xi_{k+1} & \mathcal{M}_i V \mathscr{P}_{k+1} \\ * & -\mathscr{P}_{k+1} \end{bmatrix} \leq 0 \tag{7.58}$$

where $\tilde{\mathcal{S}}_{3,k} \triangleq \sum_{l=1}^q \sum_{i=1}^N \mathscr{F}_k \Omega_k^{(l)} \mathcal{R}_i \left(\mathrm{tr}[\tilde{\Gamma}_{i,k}^{(l)} U Y_k^{-1} U^{\mathrm{T}}] + \beta_{i,k}^{(l)} \right) \mathscr{F}_k^{\mathrm{T}}$ *and the parameters* $\{X_k\}_{1 \leq k \leq T+1}$ *and* $\{Y_k\}_{1 \leq k \leq T+1}$ *are updated iteratively by*

$$\begin{cases} X_k = \mathscr{Q}_k^{-1}, \\ Y_k = \mathscr{P}_k^{-1}, \end{cases} \tag{7.59}$$

with the initial values X_0 *and* Y_0 *being set according to*

$$\begin{cases} X_0 \leq \gamma^2 U^{\mathrm{T}}(I_N \otimes \Phi) U, \\ \Xi_0 \geq \mathcal{M}_i V Y_0^{-1} V^{\mathrm{T}} \mathcal{M}_i^{\mathrm{T}}. \end{cases} \tag{7.60}$$

Proof *By noticing the parameters updating formula (7.59), we can see that the inequalities (7.54)–(7.58) are equivalent to the inequalities (7.42)–(7.46). Therefore, according to Theorem 7.3, the closed-loop nonlinear time-varying stochastic multi-agent system (7.7) reaches the desired* $(\mathscr{G}, \gamma, \Xi_k)$-*dependent consensus. The proof is now complete.*

Remark 7.4 *A sufficient condition has been proposed in Theorem 7.4 for the solvability of the addressed problem by means of the recursive linear matrix inequality approach. According to [193], the standard linear matrix inequality system has a polynomial-time complexity bounded by $\mathcal{O}(\mathcal{U}\mathcal{V}3\log(\mathcal{Z}/\varrho))$. Here, $\mathcal{U}$, $\mathcal{V}$, $\mathcal{Z}$ and ϱ represent, respectively, the total rows, the total decision variables, the data-dependant scaling coefficient and the accuracy parameter for the linear matrix inequality system. The complexity of our provided algorithm can then be easily computed by resorting to the formula developed in [193].*

Theorem 7.4 lays the principle of designing the feedback controller gains $\{K_k\}_{0 \le k \le T}$ by solving a series of recursive linear matrix inequalities. In what follows, an iterative algorithm is outlined to recursively determine K_k over the horizon $[0, T]$.

Algorithm 7.1 *Computational Algorithm for* $\{K_k\}_{0 \le k \le T}$

Step 1 Initialization: Set the time horizon T, the weighting matrix Φ and the triple $(\mathcal{G}, \gamma, \{\Xi_k\}_{0 \le k \le T})$. Set $k = 0$ and the initial values of $x_{i,0}$ satisfying (7.12). Set the values of Φ, X_0 and Y_0 satisfying (7.60). Select the value of positive scalar ϵ.

Step 2 With the known $\{\mathcal{G}, \gamma, \Xi_k, X_k, Y_k\}$, solve the set of recursive linear matrix inequalities (7.54)–(7.58) to obtain $\{K_k, \mathcal{Q}_{k+1}, \mathcal{P}_{k+1}\}$.

Step 3 Update X_{k+1} and Y_{k+1} according to (7.59) by using the obtained $\{\mathcal{Q}_{k+1}, \mathcal{P}_{k+1}\}$.

Step 4 Set $k = k + 1$. If $k > T$, exit. Otherwise, go to Step 2.

In the following, as discussed in Remark 7.3, two optimization problems are presented in terms of two corollaries to show the trade-off between the $\mathcal{H}_\infty$ and mean square requirements thereby demonstrating the flexibility of our purposed controller design method.

Corollary 7.1 *Let the pair $(\mathcal{G}, \{\Xi_k\}_{0 \le k \le T})$ and the positive-definite matrix Φ be given. Let $\epsilon > 0$ be any positive constant. The minimized disturbance attenuation level γ can be guaranteed if there exist two sequences of positive scalars $\{\alpha_{i,k}^{(l)}\}_{0 \le k \le T}$ and $\{\beta_{i,k}^{(l)}\}_{0 \le k \le T}$ $(l = 1, 2, \ldots, q; i = 1, 2, \ldots, N)$, a sequence of real-valued matrices $\{K_k\}_{0 \le k \le T}$, sequences of positive definite matrices $\{\mathcal{Q}_k\}_{1 \le k \le T+1}$ and $\{\mathcal{P}_k\}_{1 \le k \le T+1}$ solving the following optimization problem:*

$$\min_{\gamma^2, K_k, \mathcal{Q}_k, \mathcal{P}_k, \alpha_{i,k}^{(l)}, \beta_{i,k}^{(l)}} \gamma^2 \qquad (7.61)$$

$$\text{subject to} \quad (7.54) - (7.60).$$

Corollary 7.2 *Let the pair* $(\mathcal{G}, \gamma)$ *and the positive-definite matrix* Φ *be given. Let* $\epsilon > 0$ *be any positive constant. The minimized mean square criterion* Ξ_k *(in the sense of matrix trace) can be determined if there exist two sequences of positive scalars* $\{\alpha_{i,k}^{(l)}\}_{0 \leq k \leq T}$ *and* $\{\beta_{i,k}^{(l)}\}_{0 \leq k \leq T}$ $(l = 1, 2, \ldots, q; i = 1, 2, \ldots, N)$, *a sequence of real-valued matrices* $\{K_k\}_{0 \leq k \leq T}$, *sequences of positive definite matrices* $\{\mathcal{Q}_k\}_{1 \leq k \leq T+1}$ *and* $\{\mathcal{P}_k\}_{1 \leq k \leq T+1}$ *solving the following optimization problem:*

$$\min_{\Xi_{k+1}, K_k, \mathcal{Q}_k, \mathcal{P}_k, \alpha_{i,k}^{(l)}, \beta_{i,k}^{(l)}} \mathrm{tr}[\Xi_{k+1}] \qquad (7.62)$$

$$\text{subject to} \quad (7.54) - (7.60).$$

Remark 7.5 *So far, we have discussed the* $(\mathcal{G}, \gamma, \Xi_k)$*-dependent consensus control problem for a class of nonlinear discrete time-varying stochastic multi-agent system. The solvability of the addressed problem is cast into the feasibility of a set of recursive linear matrix inequalities. By means of two corollaries, it has been shown that such an algorithm is capable of making the trade-off between the two essential performance requirements (i.e., the* $\mathcal{H}_\infty$ *specification and mean square criterion) by making full use of the design flexibility.*

7.3 Illustrative Example

In this section, an illustrative example is presented to verify the effectiveness of the proposed algorithm for the addressed mean square $\mathcal{H}_\infty$ consensus control problem.

Consider the nonlinear discrete time-varying stochastic multi-agent system (7.1) with following parameters:

$$A_k = \begin{bmatrix} 0.95 + 0.10\sin(k) & -0.28 \\ 0.2 + 0.20\sin(2k) & 0.30 \end{bmatrix},$$

$$B_k = \begin{bmatrix} 0.25 - 0.15\sin(5k) \\ 0.50 \end{bmatrix},$$

$$C_k = \begin{bmatrix} 0.15 & 0.24 \end{bmatrix},$$

$$D_k = \begin{bmatrix} 0.35 + 0.20\cos(k) \\ 0.25 \end{bmatrix},$$

$$E_k = 0.20,$$

$$F_k = \begin{bmatrix} 0.25 & 0.30 \\ 0.15 & 0.40 \end{bmatrix},$$

$$L_k = \begin{bmatrix} 0.20 & 0.30 \end{bmatrix}.$$

Suppose that there are three agents connected via an undirected communication graph $\mathscr{G}$ with the associated matrix $\mathcal{H}$ as follows:

$$\mathcal{H} = \begin{bmatrix} -1 & 0 & 1 \\ 0 & -1 & 1 \\ 1 & 1 & -2 \end{bmatrix}.$$

Assume that the stochastic nonlinearities have the following form:

$$f_{i,k} = \begin{bmatrix} 0.05 \\ 0.1 \end{bmatrix} \left(0.1(x_{i,k}^{(1)}\varphi_{i,k} + x_{i,k}^{(2)}\rho_{i,k}) + 0.15u_{i,k}\theta_{i,k} \right)$$

where $\varphi_{i,k}$, $\rho_{i,k}$ and $\theta_{i,k}$ $(i = 1,2,3)$ are uncorrelated zero-mean white sequences with unitary covariances and $x_{i,k}^{(1)}$ and $x_{i,k}^{(2)}$ represent the first and second entry of the state $x_{i,k}$. Then, it can be easily checked that

$$\mathbb{E}\{f_{i,k}\} = 0,$$

$$\mathbb{E}\{f_{i,k}f_{i,k}^{\mathrm{T}}\} = \begin{bmatrix} 0.05 \\ 0.1 \end{bmatrix} \begin{bmatrix} 0.05 \\ 0.1 \end{bmatrix}^{\mathrm{T}}$$
$$\times \left(x_{i,k}^{\mathrm{T}}\mathrm{diag}\{0.01, 0.01\}x_{i,k} + 0.0225u_{i,k}^{\mathrm{T}}u_{i,k} \right).$$

Set the initial values as follows:

$$x_{1,0} = \begin{bmatrix} -2 \\ 0 \end{bmatrix},$$

$$x_{2,0} = \begin{bmatrix} -1 \\ 1 \end{bmatrix},$$

$$x_{3,0} = \begin{bmatrix} 2 \\ 1 \end{bmatrix},$$

$$X_0 = 2I_6, \quad Y_0 = 2I_{12}.$$

Set $T = 100$, $\gamma = 2$, $\Phi = I_2$ and $\Xi_k = 7I_2$. Then it can be checked that the initial conditions (7.12) and (7.60) are satisfied. With the help of Matlab software, the simulation is carried out by implementing the proposed output feedback control algorithm. The results are shown in Figs. 7.1–7.3 as well as in Table 7.1.

Fig. 7.1 plots the consensus errors $z_{i,k} - \bar{z}_k$ $(i = 1,2,3)$ which have been enforced to achieve the prespecified noise attenuation level in the $\mathcal{H}_\infty$ sense. In order to observe the mean square consensus performance, for agent i $(i = 1,2,3)$, we define $d_{i1} \triangleq (x_{i,k}^{(1)} - \bar{x}_{i,k}^{(1)})^2$ and $d_{i2} \triangleq (x_{i,k}^{(2)} - \bar{x}_{i,k}^{(2)})^2$ to characterize the real-time deviations from the two entries of agent states to their average values, respectively. The actual deviations from $x_{i,k}^{(1)}$ and $x_{i,k}^{(2)}$ to the average values are shown in Fig. 7.2 and Fig. 7.3, respectively, where we can clearly see that, at each time step, the deviation level of each individual agent is confined to be less than the prespecified upper bound. The parameters of the desired output feedback controller are listed in Table 7.1. The simulation results have confirmed that the designed mean square $\mathcal{H}_\infty$ consensus controllers perform quite well.

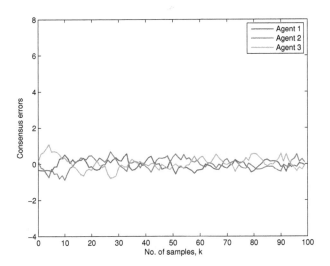

FIGURE 7.1: Consensus errors $z_{i,k} - \bar{z}_k$ $(i = 1, 2, 3)$.

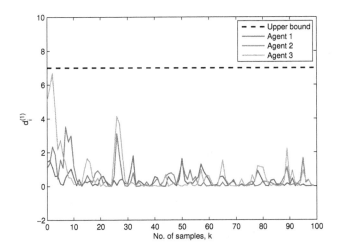

FIGURE 7.2: Mean-square performances of $x_{i,k}^{(1)}$ $(i = 1, 2, 3)$.

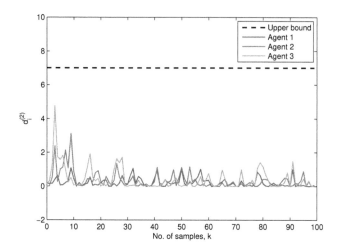

FIGURE 7.3: Mean-square performances of $x_{i,k}^{(2)}$ $(i = 1, 2, 3)$.

k	0	1	2	3	4	5
K_k	1.3072	2.0260	1.4117	2.0602	1.4882	1.4796
k	6	7	8	$\cdots$	99	100
K_k	1.9482	1.5307	1.4612	$\cdots$	1.1930	1.2396

TABLE 7.1: Output feedback controller gains

7.4   Summary

In this chapter, the mean square $\mathcal{H}_\infty$ consensus control problem has been investigated for a class of nonlinear discrete time-varying stochastic multi-agent system. The stochastic nonlinearities characterized by statistical means are quite general and could cover several well-studied nonlinearities as special cases. By means of the recursive linear matrix inequality approach, a general framework is established for the addressed multi-agent system to reach the desired consensus satisfying both $\mathcal{H}_\infty$ specification and mean square criterion. Sufficient conditions have been derived in terms of a set of recursive matrix inequalities. The effectiveness and applicability of the proposed control scheme have been verified via a numerical simulation example.

8

Consensus Control for Nonlinear Multi-Agent Systems Subject to Deception Attacks

In the past decade, multi-agent systems (MASs) have captured considerable research attention due to their extensive applications in a wide variety of areas involving chemistry manufacturing industries, geological exploration, building automation and military industries. It should be emphasized that the consensus problem for multi-agent systems has attracted special interest because of its clear practical insights. So far, many research fruits have been available in the literature and quite a lot of analysis as well as design methodologies have been exploited to deal with the consensus problems for various types of MASs.

Up to now, almost all of the multi-agent systems discussed in the literature have been assumed to be time-invariant, although a large class of engineering systems are indeed time-varying. In recent years, some interesting initial results regarding the consensus control problem with time-varying MASs have been appeared in the literature. However, despite its clear engineering significance, the consensus problem for time-varying MASs has not yet been adequately studied; this deserves our further investigation.

During the process of information transmission between the agents via the communication networks, data without security protection can be easily exploited by attackers. Network-based attacks give rise to certain challenges for those engineering applications such as consensus control for MASs since it could result in unsatisfactory consensus performance or might even cause agents to collide if not handled appropriately. Therefore, much effort has been devoted to investigation recently on the security of networks and many results have been reported. In the general context of multi-agent systems, there are mainly two types of attacks. One is the so-called Denial-of-Service (DoS) attacks, where the adversary prevents the agents from receiving information sent from their neighbors. The other is deception attacks, where the adversary replaces the true data exchanged between the agents by false signals. So far, a slice of preliminary results concerning security problems of multi-agent systems have appeared. However, when it comes to the consensus control problem for MASs, especially on the occasion when the systems under investigation are time-varying, relevant results have been very scattered. This is mainly because it is always arduous to quantify the consensus behavior in a time-varying manner, not to mention the case where cyber attacks are involved. This motivates us to shorten such a gap in the current study.

In this chapter, it is our objective to design an output-feedback controller for a class of discrete time-varying stochastic multi-agent systems subject to deception attacks. It should be pointed out that the traditional consensus results cannot be directly generalized to the time-varying case and three challenges are identified as follows: i) how can we propose a new consensus concept capable of describing the dynamical consensus characteristics for the time-varying MASs subject to the deception attacks? ii) how can we examine the impact from the given topology and the deception attacks on the consensus performance? iii) how can we design the optimal control strategy to guarantee the locally best consensus performance? We endeavor to answer the above three questions in this chapter by launching a major study on the so-called quasi-consensus control problem.

The rest of this chapter is organized as follows. Section 8.1 formulates the event-triggered output-feedback controller design problem for a discrete time-varying stochastic multi-agent system. The main results are presented in Section 8.2 where sufficient conditions for the multi-agent system reaching quasi-consensus are given in terms of iterative linear matrix inequalities. Section 8.3 outlines our conclusion.

8.1   Problem Formulation

In this chapter, the multi-agent system has N agents which communicate with each other according to a fixed network topology. The topology is represented by an undirected graph $\mathscr{G}$ consisting of a vertex set $\mathscr{V} = \{1, 2, \ldots, N\}$, an edge set $\mathscr{E} \in \mathscr{V} \times \mathscr{V}$ and a symmetric weighted adjacency matrix $\mathfrak{H} = [h_{ij}]$ with $h_{ij} \geq 0$ where $h_{ij} > 0$ if and only if $(i, j) \in \mathscr{E}$, namely, agent i can obtain information from agent j. Self-edges (i, i) are not allowed, that is, $(i, i) \notin \mathscr{E}$ for any $i \in \mathscr{V}$. Furthermore, we denote the neighborhood of agent i by $\mathscr{N}_i = \{j \in \mathscr{V} : (j, i) \in \mathscr{E}\}$. An element of $\mathscr{N}_i$ is called a neighbor of agent i. The in-degree of agent i is defined as $\deg_{\mathrm{in}}^i \triangleq \sum_{j \in \mathscr{N}_i} h_{ij}$.

Consider the following nonlinear discrete time-varying stochastic multi-agent system defined on the horizon $k \in [0, T]$:

$$\begin{cases} x_{i,k+1} = A_k x_{i,k} + B_k u_{i,k} + D_k f(x_{i,k}), \\ y_{i,k} = C_k x_{i,k} + E_k x_{i,k} v_k \end{cases} \tag{8.1}$$

where $x_{i,k} \in \mathbb{R}^n$, $y_{i,k} \in \mathbb{R}^s$ and $u_{i,k} \in \mathbb{R}^p$ are, respectively, the state, the measurement output and the control input of agent i. $v_k \in \mathbb{R}$ is a zero-mean random sequence with $\mathbb{E}\{v_k^2\} = 1$. A_k, B_k, C_k, D_k and E_k are time-varying matrices with compatible dimensions. Denote $f_{i,k} \triangleq f(x_{i,k})$. The stochastic

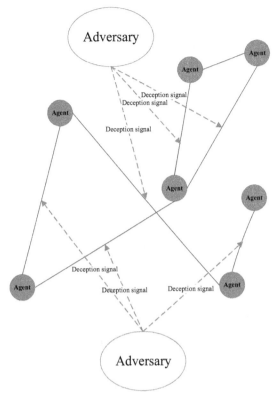

FIGURE 8.1: Deception attacks on multi-agent systems.

nonlinearities $f_{i,k}$ are assumed to have the following statistical properties:

$$\begin{aligned}
\mathbb{E}\{f_{i,k}|x_{i,k}\} &= 0, \\
\mathbb{E}\{f_{i,k}f_{l,k}^{\mathrm{T}}|x_{i,k}\} &= 0, \qquad i \neq l \\
\mathbb{E}\{f_{i,k}f_{i,k}^{\mathrm{T}}|x_{i,k}\} &\leq \sum_{j=1}^{q} \Theta_{ij,k}x_{i,k}^{\mathrm{T}}\Gamma_{ij,k}x_{i,k}
\end{aligned} \tag{8.2}$$

where $\Theta_{ij,k} \geq 0$ and $\Gamma_{ij,k} \geq 0$ are known matrices with compatible dimensions.

Remark 8.1 *The nonlinear description (8.2) can encompass several types of well-studied stochastic nonlinearities including, but not limited to, state-dependent noises (also known as multiplicative noises), nonlinearity with random sequences whose powers depend on sector bound nonlinear functions of the state and nonlinearity with random sequences whose powers depend on the sign of a nonlinear function of the state; see [134] for more details.*

In this chapter, deception cyber attacks are taken into consideration. The attack scenario is illustrated by Fig. 8.1. In order to deteriorate the consensus performance of the multi-agent system, the adversary tries to inject deception signals into the true signals of the measurements output $y_{i,k}$ during the process of data transmission between the agents via the communication network according to the given topology. It is assumed that the signals used by the adversary for the deception attacks are generated as follows:

$$\breve{y}_{i,k} = -y_{i,k} + \vartheta_{i,k}, \quad i = 1, 2, \ldots, N \tag{8.3}$$

where $\vartheta_{i,k}$ are the unknown bounded signals satisfying

$$\vartheta_{i,k}^{\mathrm{T}} R_{i,k}^{-1} \vartheta_{i,k} \leq 1, \quad i = 1, \ldots, N \tag{8.4}$$

where $R_{i,k}$ are known positive definite matrices.

It should be pointed out that the attack could not be successful each time and therefore a Bernoulli process with known statistical information has been employed to govern the DoS attacks with known success probability [4]. Similarly, from a controlled plant's viewpoint, the deception attacks may occur in a random way. As such, the measurement output can be described by

$$\begin{aligned}
\tilde{y}_{i,k} &= y_{i,k} + \beta_{i,k} \breve{y}_{i,k} \\
&= (1 - \beta_{i,k}) y_{i,k} + \beta_{i,k} \vartheta_{i,k} \\
&\triangleq \alpha_{i,k} y_{i,k} + \beta_{i,k} \vartheta_{i,k}
\end{aligned} \tag{8.5}$$

where $\alpha_{i,k} \triangleq 1 - \beta_{i,k}$, and $\beta_{i,k}$ obeys the Bernoulli distribution with

$$\mathbb{E}\{\beta_i\} = \mathcal{P}\{\beta_{i,k} = 1\} = \bar{\beta}_i, \quad \mathrm{Var}\{\beta_{i,k}\} = \sigma_i^2. \tag{8.6}$$

In this chapter, the following static output feedback controller will be adopted:

$$u_{i,k} = K_k \sum_{j \in \mathcal{N}_i} h_{ij} \left(\tilde{y}_{j,k} - y_{i,k} \right) \triangleq K_k \eta_{i,k}. \tag{8.7}$$

Then, the closed-loop dynamics of each agent can be described by

$$x_{i,k+1} = A_k x_{i,k} + B_k K_k \eta_{i,k} + D_k f_{i,k}. \tag{8.8}$$

For brevity of later development, we denote

$$x_k \triangleq \mathrm{col}_N\{x_{i,k}\}, u_k \triangleq \mathrm{col}_N\{u_{i,k}\}, \eta_k \triangleq \mathrm{col}_N\{\eta_{i,k}\},$$

$$\tilde{y}_k \triangleq \mathrm{col}_N\{\tilde{y}_{i,k}\}, \tilde{f}_k \triangleq \mathrm{col}_N\{f_{i,k}\}, \vartheta_k \triangleq \mathrm{col}_N\{\vartheta_{i,k}\},$$

$$y_k \triangleq \mathrm{col}_N\{y_{i,k}\}, \mathcal{G}_k \triangleq \mathrm{diag}_N\{\deg_{\mathrm{in}}^i\},$$

$$\mathcal{H}_k \triangleq [h_{ij}]_{N \times N} - \mathrm{diag}_N\{h_{i,i}\} - \mathcal{G}_k, \ \alpha_k \triangleq \mathrm{diag}_N\{\alpha_{i,k}\},$$

$$\beta_k \triangleq \mathrm{diag}_N\{\beta_{i,k}\}, \mathcal{Q}_k \triangleq \mathcal{H}_k + \mathcal{G}_k, \mathcal{A}_k \triangleq \mathrm{diag}_N\{A_k\},$$

$$\mathcal{K}_k \triangleq \mathrm{diag}_N\{B_k K_k\}, \mathcal{D}_k \triangleq \mathrm{diag}_N\{D_k\},$$

$$\mathcal{N}_k \triangleq I_N - \frac{1}{N} 1_N 1_N^{\mathrm{T}}, \mathscr{H}_k \triangleq \mathcal{N}_k \mathcal{H}_k, \mathscr{K}_k \triangleq B_k K_k,$$

$$\Psi_i \triangleq [\underbrace{0 \ 0 \ \cdots \ 0}_{i-1} \ I_n \ \underbrace{0 \ \cdots \ 0}_{N-i}],$$

$$\Upsilon_i \triangleq [\underbrace{0 \ 0 \ \cdots \ 0}_{i-1} \ I_l \ \underbrace{0 \ \cdots \ 0}_{N-i}],$$

$$\Sigma_i \triangleq [\underbrace{0 \ 0 \ \cdots \ 0}_{i-1} \ I_s \ \underbrace{0 \ \cdots \ 0}_{N-i}].$$

Then, the closed loop system (8.8) can be represented in a compact form as follows:

$$x_{k+1} = \mathcal{A}_k x_k + \mathcal{K}_k \eta_k + \mathcal{D}_k \tilde{f}_k. \tag{8.9}$$

Defining the average state of all agents by

$$\bar{x}_k \triangleq \mathbb{E}\left\{ \frac{1}{N} \sum_{i=1}^{N} x_{i,k} \right\} = \mathbb{E}\left\{ \frac{1}{N} (1_N^T \otimes I_n) x_k \right\} \tag{8.10}$$

and then taking (8.9) into consideration, we obtain

$$\bar{x}_{k+1} = \mathbb{E}\left\{ \frac{1}{N} (1_N^T \otimes I_n) x_{k+1} \right\}$$

$$= A_k \bar{x}_k + \frac{1}{N} (1_N^T \otimes (B_k K_k)) \eta_k. \tag{8.11}$$

Before proceeding further, we introduce the following definition and assumption.

Definition 8.1 *Let the undirected communication graph $\mathscr{G}$ and the sequence of the constraints (on the consensus performance) $\{P_k\}_{0 \leq k \leq T}$ be given. The discrete time-varying stochastic multi-agent system (8.1) is said to reach $(\mathscr{G}, P_k)$-dependent quasi-consensus if, for $i \in \mathscr{V}$ and $k \geq 0$, the following set of matrix inequalities:*

$$\Lambda_k \triangleq \mathbb{E}\left\{ (x_{i,k} - \bar{x}_k)^T P_k^{-1} (x_{i,k} - \bar{x}_k) \right\} \leq 1 \tag{8.12}$$

is satisfied.

Assumption 8.1 *The initial values $x_{i,0}$ $(i = 1, 2, \ldots, N)$ are known and satisfying*

$$(x_{i,0} - \bar{x}_0)^T P_0^{-1} (x_{i,0} - \bar{x}_0) \leq 1 \tag{8.13}$$

where $P_0 > 0$ is a given positive definite matrix.

In this chapter, for a given couple $(\mathscr{G}, \{P_k\}_{0 \leq k \leq T})$, we aim to design the controller parameter K_k in (8.7) such that the time-varying MAS (8.1) with controller (8.7) reaches the $(\mathscr{G}, P_k)$-dependent quasi-consensus despite the deception attack (8.3).

8.2 Main Results

In this section, we are going to develop a recursive scheme for determining the gain matrices K_k ensuring that the desired $(\mathscr{G}, P_k)$-dependent quasi-consensus for multi-agent system (8.1) is reached for all agents at each time instant. Before giving our main results, we first introduce the following lemmas that are useful in further development.

Lemma 8.1 *Let* $W_0(x), W_1(x), W_2(x), \dots, W_p(x)$ *be quadratic functions of* $x \in \mathbb{R}^n$, *i. e.,* $W_i(x) = x^{\mathrm{T}} Q_i x$ $(i = 0, 1, 2, \dots, p)$ *with* $Q_i^{\mathrm{T}} = Q_i$. *If there exist* $\rho_1, \rho_2, \dots, \rho_p \geq 0$ *such that* $Q_0 - \sum_{i=1}^{p} \rho_i Q_i \leq 0$, *then the following is true:*

$$W_1(x) \leq 0, \dots, W_p(x) \leq 0 \rightarrow W_0(x) \leq 0. \tag{8.14}$$

$(\mathscr{G}, P_k)$-dependent quasi-consensus

Theorem 8.1 *Given the pair* $(\mathscr{G}, \{P_k\}_{0 \leq k \leq T}))$, *for a nonlinear stochastic multi-agent system* (8.1) *subject to the randomly occurring deception attack* (8.3), *suppose the initial state* $x_{i,0}$ *satisfies* (8.13). *If there exists a sequence of real-valued matrices* $\{K_k\}_{0 \leq k \leq T}$, *sequences of positive scalars* $\{\tau_k, \varepsilon_{i,k}, \epsilon_{i,k}\}_{0 \leq k \leq T}$ $(i = 1, 2, \dots, N)$ *satisfying the following N linear matrix inequalities:*

$$\begin{bmatrix} -\Phi_k & \Xi_k^{\mathrm{T}}(I_4 \otimes \Psi_i^{\mathrm{T}}) \\ * & -\mathrm{diag}_4\{P_{k+1}\} \end{bmatrix} \leq 0 \tag{8.15}$$

where

$$\Phi_k \triangleq \mathrm{diag}\Big\{ 1 - \sum_{i=1}^{N}(\varepsilon_{i,k} + \epsilon_{i,k}), \sum_{i=1}^{N} \varepsilon_{i,k} \Upsilon_i^{\mathrm{T}} \Upsilon_i,$$

$$\sum_{i=1}^{N} \epsilon_{i,k} \Sigma_i^{\mathrm{T}} R_{i,k}^{-1} \Sigma_i, \tau_k I \Big\} - \tau_k \Omega_k^{\mathrm{T}} \Omega_k, \tag{8.16}$$

$$\bar{\Pi}_k^{(1,1)} \triangleq \big((\mathscr{H}_k \bar{\alpha} \mathbf{1}_N) \otimes (\mathscr{K}_k C_k) \big) \bar{x}_k,$$

$$\bar{\Pi}_k^{(1,2)} \triangleq I_N \otimes (A_k F_k) + (\mathscr{H}_k \bar{\alpha}) \otimes (\mathscr{K}_k C_k F_k),$$

$$\bar{\Pi}_k^{(1,3)} \triangleq (\mathscr{H}_k \bar{\beta}) \otimes \mathscr{K}_k,$$

$$\bar{\Pi}_{v,k}^{(1,1)} \triangleq \big((\mathscr{H}_k \bar{\alpha} \mathbf{1}_N) \otimes (\mathscr{K}_k E_k) \big) \bar{x}_k,$$

$$\bar{\Pi}_{v,k}^{(1,2)} \triangleq (\mathscr{H}_k \bar{\alpha}) \otimes (\mathscr{K}_k E_k F_k),$$

$$\hat{\Pi}_k^{(1,1)} \triangleq \big((\mathscr{H}_k \sigma \mathbf{1}_N) \otimes (\mathscr{K}_k C_k) \big) \bar{x}_k,$$

$$\hat{\Pi}_k^{(1,2)} \triangleq (\mathscr{H}_k \sigma) \otimes (\mathscr{K}_k C_k F_k),$$

$$\check{\Pi}_{v,k}^{(1,1)} \triangleq \big((\mathscr{H}_k \sigma \mathbf{1}_N) \otimes (\mathscr{K}_k E_k) \big) \bar{x}_k,$$

$$\check{\Pi}_{v,k}^{(1,2)} \triangleq (\mathscr{H}_k \sigma) \otimes (\mathscr{K}_k E_k F_k),$$

$$\bar{\Pi}_k \triangleq \begin{bmatrix} \bar{\Pi}_k^{(1,1)} & \bar{\Pi}_k^{(1,2)} & \bar{\Pi}_k^{(1,3)} & \mathcal{D}_k \end{bmatrix}, \tag{8.17}$$

$$\bar{\Pi}_{v,k} \triangleq \begin{bmatrix} \bar{\Pi}_{v,k}^{(1,1)} & \bar{\Pi}_{v,k}^{(1,2)} & 0 & 0 \end{bmatrix}, \tag{8.18}$$

$$\hat{\Pi}_k \triangleq \begin{bmatrix} \hat{\Pi}_k^{(1,1)} & \hat{\Pi}_k^{(1,2)} & (\mathscr{H}_k \sigma) \otimes \mathscr{K}_k & 0 \end{bmatrix}, \tag{8.19}$$

$$\check{\Pi}_{v,k} \triangleq \begin{bmatrix} \check{\Pi}_{v,k}^{(1,1)} & \check{\Pi}_{v,k}^{(1,2)} & 0 & 0 \end{bmatrix}, \tag{8.20}$$

$$\Xi_k^{\mathrm{T}} \triangleq \begin{bmatrix} \bar{\Pi}_k^{\mathrm{T}} & \bar{\Pi}_{v,k}^{\mathrm{T}} & \hat{\Pi}_k^{\mathrm{T}} & \check{\Pi}_{v,k}^{\mathrm{T}} \end{bmatrix}. \tag{8.21}$$

then, the multi-agent system (8.1) reaches the desired quasi-consensus.

Proof *First of all, noting (8.7), we acquire*

$$\eta_{i,k} = \sum_{j \in \mathcal{N}_i} h_{ij} \left(\tilde{y}_{j,k} - y_{i,k} \right)$$

$$= \sum_{j \in \mathcal{N}_i} h_{ij} \left(\tilde{y}_{j,k} - \tilde{y}_{i,k} + \tilde{y}_{i,k} - y_{i,k} \right)$$

$$= \sum_{j \in \mathcal{N}_i} h_{ij} \left(\tilde{y}_{j,k} - \tilde{y}_{i,k} \right) + \sum_{j \in \mathcal{N}_i} h_{ij} \left(\tilde{y}_{i,k} - y_{i,k} \right), \qquad (8.22)$$

or, equivalently,

$$\begin{aligned}
\eta_k =& (\mathcal{H}_k \otimes I_m) \tilde{y}_k - \left((\mathcal{G}_k \beta_k) \otimes I_m \right) y_k \\
&+ \left((\mathcal{G}_k \beta_k) \otimes I_m \right) \vartheta_k \\
=& \left((\mathcal{H}_k - \mathcal{Q}_k \beta_k) \otimes C_k \right) x_k \\
&+ \left((\mathcal{H}_k - \mathcal{Q}_k \beta_k) \otimes E_k \right) x_k v_k \\
&+ \left((\mathcal{Q}_k \beta_k) \otimes I_m \right) \vartheta_k.
\end{aligned} \qquad (8.23)$$

Substituting (8.23) into (9.9) yields

$$\begin{aligned}
x_{k+1} =& \mathcal{A}_k x_k + \mathcal{K}_k \eta_k + \mathcal{D}_k \tilde{f}_k \\
=& \left(\mathcal{A}_k + (\mathcal{H}_k \alpha_k) \otimes (B_k K_k C_k) \right) x_k \\
&+ \left((\mathcal{H}_k \alpha_k) \otimes (B_k K_k E_k) \right) x_k v_k \\
&+ \left((\mathcal{H}_k \beta_k) \otimes (B_k K_k) \right) \vartheta_k + \mathcal{D}_k \tilde{f}_k.
\end{aligned} \qquad (8.24)$$

In order to look at the deviation of each agent's state from the average state, we denote $\tilde{x}_{i,k} \triangleq x_{i,k} - \bar{x}_k$ and $\tilde{x}_k \triangleq [\tilde{x}_{1,k}^{\mathrm{T}} \; \tilde{x}_{2,k}^{\mathrm{T}} \cdots \tilde{x}_{N,k}^{\mathrm{T}}]^{\mathrm{T}}$. Then, subtracting (8.11) from (8.1) leads to

$$\begin{aligned}
\tilde{x}_{i,k+1} =& x_{i,k+1} - \bar{x}_{k+1} \\
=& A_k x_{i,k} + B_k u_{i,k} + D_k f_{i,k} \\
&- \left(A_k \bar{x}_k + \frac{1}{N} \left(\mathbf{1}_N^{\mathrm{T}} \otimes (B_k K_k) \right) \eta_k \right) \\
=& A_k x_{i,k} + B_k K_k \eta_{i,k} + D_k f_{i,k} \\
&- \left(A_k \bar{x}_k + \frac{1}{N} \left(\mathbf{1}_N^{\mathrm{T}} \otimes (B_k K_k) \right) \eta_k \right).
\end{aligned} \qquad (8.25)$$

Taking (8.24) into consideration, we can easily have

$$\begin{aligned}
\tilde{x}_{k+1} =& x_{k+1} - (\mathbf{1}_N \otimes I_n) \bar{x}_{k+1} \\
=& \left(\mathcal{A}_k + (\mathcal{N}_k \mathcal{H}_k \alpha_k) \otimes (B_k K_k C_k) \right) x_k + \mathcal{D}_k \tilde{f}_k \\
&+ \left((\mathcal{N}_k \mathcal{H}_k \alpha_k) \otimes (B_k K_k E_k) \right) x_k v_k \\
&+ \left((\mathcal{N}_k \mathcal{H}_k \beta_k) \otimes (B_k K_k) \right) \vartheta_k \\
&- (\mathbf{1}_N \otimes A_k) \bar{x}_k.
\end{aligned} \qquad (8.26)$$

Then, $\tilde{x}_{k+1}$ can be rewritten as

$$
\begin{aligned}
\tilde{x}_{k+1} =& x_{k+1} - (\mathbf{1}_N \otimes I_n)\bar{x}_{k+1} \\
=& \big(\mathcal{A}_k + (\mathscr{H}_k \alpha_k) \otimes (\mathscr{K}_k C_k)\big)x_k + \mathcal{D}_k \tilde{f}_k \\
& + \big((\mathscr{H}_k \alpha_k) \otimes (\mathscr{K}_k E_k)\big)x_k v_k \\
& + \big((\mathscr{H}_k \beta_k) \otimes \mathscr{K}_k\big)\vartheta_k - (\mathbf{1}_N \otimes A_k)\bar{x}_k.
\end{aligned}
\tag{8.27}
$$

In the following, we proceed to prove the theorem by virtue of induction. First, it can be known from Assumption 8.1 that $\Lambda_0 \le 1$ holds. Then, suppose that $\Lambda_k \le 1$ is true at time instant $k > 0$; we are now going to prove that $\Lambda_{k+1} \le 1$ holds with the condition given in this theorem.

According to [18], if

$$
\mathbb{E}\{(x_{i,k} - \bar{x}_k)^{\mathrm{T}} P_k^{-1}(x_{i,k} - \bar{x}_k)\} \le 1,
\tag{8.28}
$$

then there exists a $z_{i,k} \in \mathbb{R}^l$ with $\mathbb{E}\{\|z_{i,k}\|\} \le 1$ such that

$$
x_{i,k} = \bar{x}_k + F_k z_{i,k}
\tag{8.29}
$$

where F_k is a factorization of $P_k = F_k F_k^{\mathrm{T}}$. Then, it is evident that x_k can be described by

$$
x_k = (\mathbf{1}_N \otimes I_n)\bar{x}_k + (I_N \otimes F_k)z_k
\tag{8.30}
$$

where $z_k \triangleq \mathrm{col}_N\{z_{i,k}\}$.

By using (8.29) and (8.30), (8.27) can be rewritten as

$$
\begin{aligned}
\tilde{x}_{k+1} =& (\mathbf{1}_N \otimes A_k)\bar{x}_k + \big(I_N \otimes (A_k F_k)\big)z_k \\
& + \big((\mathscr{H}_k \alpha_k \mathbf{1}_N) \otimes (\mathscr{K}_k C_k)\big)\bar{x}_k \\
& + \big((\mathscr{H}_k \alpha_k) \otimes (\mathscr{K}_k C_k F_k)\big)z_k \\
& + \big((\mathscr{H}_k \alpha_k \mathbf{1}_N) \otimes (\mathscr{K}_k E_k)\big)v_k \bar{x}_k \\
& + \big((\mathscr{H}_k \alpha_k) \otimes (\mathscr{K}_k E_k F_k)\big)v_k z_k \\
& + \big((\mathscr{H}_k \beta_k) \otimes \mathscr{K}_k\big)\vartheta_k \\
& - (\mathbf{1}_N \otimes A_k)\bar{x}_k + \mathcal{D}_k \tilde{f}_k \\
=& \mathfrak{A}_k \bar{x}_k + \mathfrak{B}_k z_k + \mathfrak{C}_k \vartheta_k + \mathcal{D}_k \tilde{f}_k
\end{aligned}
\tag{8.31}
$$

where

$$
\begin{aligned}
\mathfrak{A}_k \triangleq& (\mathscr{H}_k \alpha_k \mathbf{1}_N) \otimes (\mathscr{K}_k C_k) + \big((\mathscr{H}_k \alpha_k \mathbf{1}_N) \otimes (\mathscr{K}_k E_k)\big)v_k, \\
\mathfrak{B}_k \triangleq& I_N \otimes (A_k F_k) + \big((\mathscr{H}_k \alpha_k) \otimes (\mathscr{K}_k C_k F_k)\big) \\
& + \big((\mathscr{H}_k \alpha_k) \otimes (\mathscr{K}_k E_k F_k)\big)v_k, \\
\mathfrak{C}_k \triangleq& (\mathscr{H}_k \beta_k) \otimes \mathscr{K}_k.
\end{aligned}
$$

Denoting

$$\xi_k \triangleq \begin{bmatrix} 1 \\ z_k \\ \vartheta_k \\ \tilde{f}_k \end{bmatrix} \tag{8.32}$$

and

$$\Pi_k \triangleq \begin{bmatrix} \mathfrak{A}_k \bar{x}_k & \mathfrak{B}_k & \mathfrak{C}_k & \mathfrak{D}_k \end{bmatrix}, \tag{8.33}$$

we have

$$\tilde{x}_{k+1} = \Pi_k \xi_k. \tag{8.34}$$

Note that there exist stochastic variables α_k, β_k and v_k in (8.33). In the following stage, for the convenience of later derivation, we denote

$$\tilde{\beta}_k \triangleq \beta_k - \bar{\beta}, \qquad \tilde{\alpha}_k \triangleq \alpha_k - \bar{\alpha} = -\tilde{\beta}_k \tag{8.35}$$

where $\bar{\beta} \triangleq \mathrm{diag}_N\{\bar{\beta}_i\}$ and $\bar{\alpha} \triangleq \mathrm{diag}_N\{1 - \bar{\beta}_i\}$. Subsequently, it is not difficult to obtain the variances of $\tilde{\alpha}_k$ and $\tilde{\beta}_k$ as follows:

$$\begin{aligned} \mathbb{E}\{\tilde{\alpha}_k^{\mathrm{T}} \tilde{\alpha}_k\} &= \mathbb{E}\{(\alpha_k - \bar{\alpha})^{\mathrm{T}}(\alpha_k - \bar{\alpha})\} \\ &= \mathbb{E}\{(-\tilde{\beta}_k)^{\mathrm{T}}(-\tilde{\beta}_k)\} \\ &= \mathbb{E}\{(\beta_k - \bar{\beta})^{\mathrm{T}}(\beta_k - \bar{\beta})\} \\ &= \mathrm{diag}\{\sigma_1^2, \sigma_2^2, \ldots, \sigma_N^2\} \\ &\triangleq \sigma^2. \end{aligned} \tag{8.36}$$

Decompose the stochastic matrix Π_k by

$$\Pi_k = \bar{\Pi}_k + \bar{\Pi}_{v,k} v_k + \tilde{\Pi}_k + \tilde{\Pi}_{v,k} v_k \tag{8.37}$$

where $\bar{\Pi}_k$ and $\bar{\Pi}_{v,k}$ are defined in (8.17)–(8.18), and $\tilde{\Pi}_k$ and $\tilde{\Pi}_{v,k}$ are defined as follows:

$$\begin{aligned} \tilde{\Pi}_k^{(1,1)} &\triangleq \big((\mathscr{H}_k \tilde{\alpha}_k \mathbf{1}_N) \otimes (\mathscr{K}_k C_k)\big) \bar{x}_k, \\ \tilde{\Pi}_k^{(1,2)} &\triangleq (\mathscr{H}_k \tilde{\alpha}_k) \otimes (\mathscr{K}_k C_k F_k), \\ \tilde{\Pi}_{v,k}^{(1,1)} &\triangleq \big((\mathscr{H}_k \tilde{\alpha}_k \mathbf{1}_N) \otimes (\mathscr{K}_k E_k)\big) \bar{x}_k, \\ \tilde{\Pi}_{v,k}^{(1,2)} &\triangleq (\mathscr{H}_k \tilde{\alpha}_k) \otimes (\mathscr{K}_k E_k F_k), \\ \tilde{\Pi}_k &\triangleq \begin{bmatrix} \tilde{\Pi}_k^{(1,1)} & \tilde{\Pi}_k^{(1,2)} & (\mathscr{H}_k \tilde{\beta}_k) \otimes \mathscr{K}_k & 0 \end{bmatrix}, \\ \tilde{\Pi}_{v,k} &\triangleq \begin{bmatrix} \tilde{\Pi}_{v,k}^{(1,1)} & \tilde{\Pi}_{v,k}^{(1,2)} & 0 & 0 \end{bmatrix}. \end{aligned}$$

Then, by taking into account the statistical properties of random variables β_k and v_k, we can find out that the following derivation is true:

$$\begin{aligned} &\mathbb{E}\left\{\xi_k^{\mathrm{T}} \Pi_k^{\mathrm{T}} \Psi_i^{\mathrm{T}} P_{k+1}^{-1} \Psi_i \Pi_k \xi_k\right\} \\ &= \xi_k^{\mathrm{T}} \left(\Xi_k^{\mathrm{T}} \big(I_4 \otimes (\Psi_i^{\mathrm{T}} P_{k+1}^{-1} \Psi_i)\big) \Xi_k\right) \xi_k \end{aligned} \tag{8.38}$$

where Ξ_k is defined in (8.21).

From (8.29) and (8.4), we have

$$\mathbb{E}\{\|z_{i,k}\|\} \leq 1, \quad \vartheta_{i,k}^{\mathrm{T}} R_k^{-1} \vartheta_{i,k} \leq 1 \tag{8.39}$$

which can be expressed by the vector ξ_k as follows:

$$\mathbb{E}\left\{\xi_k^{\mathrm{T}} \mathscr{U}_i \xi_k\right\} \leq 0, \quad \xi_k^{\mathrm{T}} \mathscr{W}_i \xi_k \leq 0. \tag{8.40}$$

where

$$\mathscr{U}_i \triangleq \mathrm{diag}\left\{-1, \Upsilon_i^{\mathrm{T}} \Upsilon_i, 0, 0\right\},$$
$$\mathscr{W}_i \triangleq \mathrm{diag}\left\{-1, 0, \Sigma_i^{\mathrm{T}} R_{i,k}^{-1} \Sigma_i, 0\right\}.$$

Similarly, denoting

$$\tilde{\Gamma}_k \triangleq \mathrm{diag}_N\left\{\sum_{j=1}^{q} \Gamma_{ij} \mathrm{tr}[\Theta_{ij}]\right\}$$

and then taking the statistical property of $f_{i,k}$ in (8.2) into consideration, we obtain

$$\mathbb{E}\{\tilde{f}_k^{\mathrm{T}} \tilde{f}_k\} = \mathbb{E}\{\mathrm{tr}[\tilde{f}_k \tilde{f}_k^{\mathrm{T}}]\}$$
$$\leq \mathbb{E}\{x_k^{\mathrm{T}} \tilde{\Gamma}_k x_k\} = \mathbb{E}\{\xi_k^{\mathrm{T}} \Omega_k^{\mathrm{T}} \Omega_k \xi_k\} \tag{8.41}$$

where $\Omega_k \triangleq \begin{bmatrix} \bar{\Gamma}_k(\mathbf{1}_N \otimes I_n)\bar{x}_k & \bar{\Gamma}_k(I_N \otimes F_k) & 0 & 0 \end{bmatrix}$ with $\bar{\Gamma}_k$ being a factorization of $\tilde{\Gamma}_k = \bar{\Gamma}_k^{\mathrm{T}} \bar{\Gamma}_k$. Consequently, it can be readily checked that (8.41) can be expressed by ξ_k as follows:

$$\mathbb{E}\left\{\xi_k^{\mathrm{T}}\left(\mathrm{diag}\{0,0,0,I\} - \Omega_k^{\mathrm{T}} \Omega_k\right)\xi_k\right\} \leq 0. \tag{8.42}$$

According to the induction principle, it remains to show that $\Lambda_{k+1} \leq 1$ holds if the conditions of this theorem are true. To this end, by utilizing the Schur Complement Lemma [18] to (8.15), we have

$$-\Phi_k + \bar{\Pi}_k^{\mathrm{T}} \Psi_i^{\mathrm{T}} P_{k+1}^{-1} \Psi_i \bar{\Pi}_k + \bar{\Pi}_{v,k}^{\mathrm{T}} \Psi_i^{\mathrm{T}} P_{k+1}^{-1} \Psi_i \bar{\Pi}_{v,k}$$
$$+ \hat{\Pi}_k^{\mathrm{T}} \Psi_i^{\mathrm{T}} P_{k+1}^{-1} \Psi_i \hat{\Pi}_k + \check{\Pi}_{v,k}^{\mathrm{T}} \Psi_i^{\mathrm{T}} P_{k+1}^{-1} \Psi_i \check{\Pi}_{v,k} \leq 0, \tag{8.43}$$

or, equivalently,

$$-\Phi_k + \Xi_k^{\mathrm{T}}\left(I_4 \otimes (\Psi_i^{\mathrm{T}} P_{k+1}^{-1} \Psi_i)\right)\Xi_k \leq 0. \tag{8.44}$$

Noticing (8.16), we obtain

$$\Xi_k^{\mathrm{T}}\left(I_4 \otimes (\Psi_i^{\mathrm{T}} P_{k+1}^{-1} \Psi_i)\right)\Xi_k - \mathrm{diag}\{1,0,0,0\}$$
$$-\sum_{i=1}^{N} \varepsilon_{i,k} \mathscr{U}_i - \sum_{i=1}^{N} \epsilon_{i,k} \mathscr{W}_i$$
$$-\tau_k\left(\mathrm{diag}\{0,0,0,I\} - \Omega_k^{\mathrm{T}} \Omega_k\right) \leq 0. \tag{8.45}$$

It follows readily from Lemma 8.1 that

$$\xi_k\Big(\Xi_k^{\mathrm{T}}\big(I_4\otimes(\Psi_i^{\mathrm{T}}P_{k+1}^{-1}\Psi_i)\big)\Xi_k - \mathrm{diag}\{1,0,0,0\}\Big)\xi_k \le 0, \tag{8.46}$$

which, by (8.38), further indicates that

$$\mathbb{E}\Big\{\xi_k^{\mathrm{T}}\Pi_k^{\mathrm{T}}\Psi_i^{\mathrm{T}}P_{k+1}^{-1}\Psi_i\Pi_k\xi_k$$
$$- \xi_k^{\mathrm{T}}\mathrm{diag}\{1,0,0,0\}\xi_k\Big\} \le 0. \tag{8.47}$$

Consequently, $\mathbb{E}\big\{\tilde{x}_{k+1}^{\mathrm{T}}\Psi_i^{\mathrm{T}}P_{k+1}^{-1}\Psi_i\tilde{x}_{k+1}\big\}\le 1$ *holds, which means the following inequality is true:*

$$\mathbb{E}\big\{(x_{i,k+1}-\bar{x}_{k+1})^{\mathrm{T}}P_{k+1}^{-1}(x_{i,k+1}-\bar{x}_{k+1})\big\}\le 1. \tag{8.48}$$

Therefore, $\Lambda_{k+1}\le 1$ *holds, and the induction is accomplished. It can now be concluded from Definition 8.1 that the multi-agent system (8.1) reaches the desired* $(\mathscr{G},P_k)$*-dependent quasi-consensus. The proof is now complete.*

Optimization problem

In this subsection, an optimization problem is formulated to seek the locally optimal consensus performance by minimizing P_k (in the sense of matrix trace).

Corollary 8.1 *For the discrete time-varying multi-agent system (8.1), let the communication graph* $\mathscr{G}$ *be given. The minimized* P_k *can be guaranteed (in the sense of the matrix trace) if there exists a sequence of real-valued matrices* $\{K_k\}_{0\le k\le T}$*, sequences of positive scalars* $\{\tau_k,\varepsilon_{i,k},\epsilon_{i,k}\}_{0\le k\le T}$ $(i=1,2,\dots,N)$ *solving the following optimization problem:*

$$\min_{P_{k+1},K_k,\tau_k,\varepsilon_{i,k},\epsilon_{i,k}} \mathrm{tr}[P_{k+1}] \tag{8.49}$$

$$\text{subject to} \quad (8.15).$$

Remark 8.2 *So far, we have discussed the* $(\mathscr{G},P_k)$*-dependent quasi-consensus problem for a discrete time-varying stochastic multi-agent system. The solvability of the addressed problem is cast into the feasibility of a set of recursive matrix inequalities. In addition, an optimization problem has been investigated with the purpose of seeking the locally optimal consensus performance.*

8.3   Summary

In this chapter, the quasi-consensus control problem has been investigated for a class of discrete time-varying nonlinear stochastic multi-agent systems subject to deception attacks. A new definition of quasi-consensus has been first presented to characterize the consensus process where all the agents are constrained to stay within a certain ellipsoidal region at each time step. A novel deception attack model has been proposed, where the attack signals are injected by the adversary into the measurement data during the process of information transmission via the communication network. By resorting to the recursive matrix inequality approach, sufficient conditions have been established for the solvability of the quasi-consensus control problem. Subsequently, an optimization problem has been provided to determine the feedback gain parameters that guarantee the locally optimal consensus performance.

9

Distributed Event-Based Set-Membership
Filtering for A Class of Nonlinear Systems
with Sensor Saturations over Sensor
Networks

The past decades have witnessed rapid growth in the utilization of sensor networks consisting of a large number of sensing nodes geographically distributed
in certain areas. Sensor networks have found extensive applications in various fields ranging from information collection, environmental monitoring, and
industrial automation to intelligent buildings. The practical significance of
sensor networks has recently led to considerable research interest in the distributed estimation or filtering problems whose aim is to extract true signals
based on the information measurements collected/transmitted via sensor networks. Compared with traditional filtering algorithms in a single sensor system, the key feature of distributed filtering over sensor networks is that each
sensor estimates the system state based not only on its own measurement but
also on the neighboring sensors' measurements according to the topology. So
far, much effort has been made to investigate the distributed filtering problems
and several effective strategies have been developed. It is worth mentioning
that, up to now, the resource efficiency issue has not been adequately addressed in relation to distributed filtering problems especially for nonlinear
time-varying systems, and this gives rise to the primary motivation of our
current research.

With today's revolution of microelectronics techniques, there is an incre-
mental adoption of small-size micro-processors which are embedded in the
sensing nodes responsible for information collecting, signal processing, data
transmitting and sometimes instruction actuating within the sensor networks.
In engineering practice, these micro-processors are apparently subject to limited resources such as battery storage. For energy-saving purposes, it is often
favorable to exploit the event-based rules under which the information received by sensing nodes is transmitted to the controllers/filters only when some
events occur. Compared to the traditional time-based communication mechanism, the event-based communication scheme has the advantage of improving
the efficiency of resource utilization by reducing the unnecessary executions
over the network. Due to its clear physical implication and promising application prospect, in the past decade or so, the event-based filtering problem

has stirred remarkable interest and many research results have been reported in the literature. It is worth noting that, despite the recent progress in event-triggering filter/control, it remains an open problem to develop more generic triggering conditions that could result in a reasonable tradeoff between the efficiency of resource utilization and the specification of system performance.

Apart from the aforementioned resource limitation issue, it is well known that embedded micro-processors are typically of limited capacity within a sensor network due primarily to physical and communication constraints. Consequently, some new phenomena (e.g. signal quantization, sensor saturation and actuator failures) have inevitably emerged that deserve particular attention in system design. These phenomena are customarily referred to as the incomplete information that has attracted much research interest in developing filtering schemes. However, when it comes to event-based distributed filtering problems with incomplete information, the corresponding results have been very few owing mainly to the lack of appropriate techniques for coping with 1) the complicated node coupling according to the topological information and 2) the demanding triggering mechanism accounting for limited capability. As such, another motivation for our current investigation is to examine the impact of incomplete information on the performance of event-based distributed filtering over the sensor network with a given topology.

On another research frontier, the set-membership filtering problem originated in [224] aims to use measurements to calculate recursively a bounding ellipsoid into the set of possible states. Recently, there has been renewed interest in set-membership filtering problems for various systems by developing computationally efficient algorithms. Unfortunately, for large-scale distributed systems such as sensor networks, set-membership filtering has not received adequate research attention, and this motivates us to investigate the set-membership filtering problem for nonlinear systems under an event-based distributed information processing mechanism.

Motivated by the above discussions, in this chapter, it is our objective to design a distributed event-triggering set-membership filtering scheme for a class of discrete time-varying nonlinear systems subject to unknown but bounded noises and sensor saturations. A novel triggering condition with clear engineering insight is proposed to better reflect the reality of practical applications. Nonlinearity is assumed to satisfy the so-called sector condition, which is quite general and could cover several classes of nonlinearities as special cases. *We endeavor to answer the following questions: i) how to deal with the proposed triggering condition within the unified framework for filter analysis and synthesis; ii) how to quantify the influences on the filtering performance from the given topology, the sector-bounded nonlinearity, the unknown but bounded noises as well as the sensor saturations; iii) how to characterize the relationship between the triggering threshold and the filtering performance or, in other words, how to exploit the trade-offs between the size of the ellipsoids and the triggering threshold so as to make compromise between the filtering performance and the triggering frequency.* We shall respond to the three questions raised above by investigating the so-called distributed set-membership filtering problem.

The rest of this chapter is organized as follows. Section 9.1 formulates the distributed event-based filter design problem for a nonlinear discrete time-varying system with unknown but bounded noises as well as sensor saturations. Our main results are presented in Section 9.2 where sufficient conditions for the existence of the desired filter are given in terms of recursive linear matrix inequalities (RLMIs). Section 9.3 gives a numerical example and Section 9.4 draws our conclusion.

9.1 Problem Formulation

In this chapter, it is assumed that the sensor network has N sensor nodes which are distributed in the space according to a specific interconnection topology characterized by a directed graph $\mathscr{G} = (\mathscr{V}, \mathscr{E}, \mathscr{L})$, where $\mathscr{V} = \{1, 2, ..., N\}$ denotes the set of sensing nodes, $\mathscr{E} \subseteq \mathscr{V} \times \mathscr{V}$ is the set of edges, and $\mathscr{L} = [\theta_{ij}]_{N \times N}$ is the nonnegative adjacency matrix associated with the edges of the graph; that is, $\theta_{ij} > 0$ if and only if edge$(i, j) \in \mathscr{E}$ (i.e. there is information transmission from sensor j to sensor i). If $(i, j) \in \mathscr{E}$, then node j is called one of the neighbors of node i. Also, we assume that $\theta_{ii} = 1$ for all $i \in \mathscr{V}$ and, therefore, (i, i) can be regarded as an additional edge. The set of neighbors of node $i \in \mathscr{V}$ plus the node itself is denoted by $\mathscr{N}_i \triangleq \{j \in \mathscr{V} | (i, j) \in \mathscr{E}\}$.

Consider a time-varying nonlinear system with N sensors described by the following state-space model:

$$\begin{cases} x_{k+1} = A_k x_k + D_k w_k + f(x_k) \\ y_{i,k} = \sigma(C_{i,k} x_k) + E_{i,k} v_k \quad i = 1, 2, \ldots, N \end{cases} \tag{9.1}$$

where $x_k \in \mathbb{R}^n$ is the system state and $y_{i,k} \in \mathbb{R}^m$ is the measurement output measured by sensor i. The parameters A_k, D_k, $C_{i,k}$ and $E_{i,k}$ are real-valued time-varying matrices of appropriate dimensions. $w_k \in \mathbb{R}^\omega$ and $v_k \in \mathbb{R}^\nu$ represent the process and measurement noises, respectively, which are deterministic and satisfy the following assumption.

Assumption 9.1 *The noise sequences w_k and v_k are confined to the following ellipsoidal sets:*

$$\begin{cases} \mathscr{S}_k \triangleq \{w_k : w_k^{\mathrm{T}} S_k^{-1} w_k \le 1\} \\ \mathscr{R}_k \triangleq \{v_k : v_k^{\mathrm{T}} R_k^{-1} v_k \le 1\} \end{cases} \tag{9.2}$$

where $S_k > 0$ and $R_k > 0$ are known matrices with compatible dimensions characterizing the sizes and orientations of the ellipsoids.

Remark 9.1 *In practical engineering, due to man-made electromagnetic interference as well as other natural sources, sometimes the noises are not really* *stochastic. Rather, they are deterministic, unknown but bounded (by energy or amplitude) [66]. As such, most statistics-based filtering algorithms (such as the Kalman filtering scheme requiring exact information on the Gaussian noises) are no longer applicable. It is worth noting that the unknown but bounded noise serves as an important type of non-Gaussian noise that has received considerable research attention with respect to the filtering problems, see e.g. [66, 201]. In this chapter, the process noises w_k and measurement noises v_k are assumed to be deterministic, unknown but bounded within certain ellipsoidal sets, and this gives rise to the set-membership filtering problem to be addressed in the sequel.*

Definition 9.1 *[103] Let K_1 and K_2 be some real matrices with $K \triangleq K_2 - K_1 > 0$. A nonlinearity $\kappa(\cdot)$ is said to satisfy the sector condition with respect to K_1 and K_2 if*

$$\big(\kappa(y) - K_1 y\big)^{\mathrm{T}} \big(\kappa(y) - K_2 y\big) \leq 0. \tag{9.3}$$

In this case, the sector-bounded nonlinearity $\kappa(\cdot)$ is said to belong to the sector $[K_1, K_2]$.

Assumption 9.2 *The nonlinear function $f(x_k)$ in the system (9.1) belongs to the sector $[U_1, U_2]$, where U_1 and U_2 are known real-valued matrices with appropriate dimensions.*

The saturation function $\sigma(\cdot)$ is defined as

$$\sigma(\cdot) = \begin{bmatrix} \sigma_1(y^{(1)}) & \sigma_2(y^{(2)}) & \cdots & \sigma_m(y^{(m)}) \end{bmatrix} \tag{9.4}$$

where $\sigma_s(y^{(s)}) = \mathrm{sign}(y^{(s)}) \min\{y_{\max}^{(s)}, |y^{(s)}|\}$ with $y^{(s)}$ representing the sth entry of the vector y. Note that, if there exist diagonal matrices G_{1i} and G_{2i} such that $0 \leq G_{1i} < I \leq G_{2i}$, then the saturation function $\sigma(C_{i,k} x_k)$ in (9.1) can be written as follows:

$$\sigma(C_{i,k} x_k) = G_{1i} C_{i,k} x_k + \varphi(C_{i,k} x_k) \tag{9.5}$$

where $\varphi(C_{i,k} x_k)$ is a certain nonlinear vector-valued function satisfying the sector condition with $K_1 = 0$ and $K_2 = G_i \triangleq G_{2i} - G_{1i}$; that is, $\varphi(C_{i,k} x_k)$ satisfies the following inequality:

$$\varphi^{\mathrm{T}}(C_{i,k} x_k) \big(\varphi(C_{i,k} x_k) - G_i C_{i,k} x_k\big) \leq 0. \tag{9.6}$$

Before introducing the distributed *event-based* filter structure, we first recall the traditional distributed time-based filter as follows:

$$\hat{x}_{i,k+1} = F_{i,k} \hat{x}_{i,k} + \sum_{j \in \mathcal{N}_i} \theta_{ij} H_{ij,k} r_{j,k} \quad i = 1, 2, \dots, N \tag{9.7}$$

where $\hat{x}_{i,k} \in \mathbb{R}^n$ is the estimate of the system state based on the ith sensing node, $F_{i,k}$ and $H_{ij,k}$ are the filter parameters, and $r_{i,k}$ represents the innovation sequence defined by

$$r_{i,k} \triangleq y_{i,k} - C_{i,k} \hat{x}_{i,k}. \tag{9.8}$$

Remark 9.2 *In traditional distributed filtering algorithms, it is usually assumed that the sensing nodes broadcast their local information at every periodic sampling instant, and this might result in unnecessary waste of communication resources especially when energy saving becomes a concern. For the purpose of improving the efficiency of network utilization, as an alternative to the periodic control method, the event-triggering mechanism will be proposed here to reduce the network communication burden with guaranteed filtering performance, where the main idea is to broadcast important messages rather than all messages.*

Let us now elaborate upon the event-triggering mechanism to be adopted. Suppose that the sequence of the triggering instants is $\{k_t^i\}$ $(t = 0, 1, 2, \ldots)$ satisfying $0 < k_0^i < k_1^i < k_2^i < \cdots < k_t^i < \cdots$, where k_t^i represents the time instant k at which the $(t+1)$th trigger occurs for agent i. Then, define

$$e_{i,k} \triangleq r_{i,k_t^i} - r_{i,k} \tag{9.9}$$

which indicates the difference between the broadcast innovations at the latest triggering time and the current time. With the notation of $e_{i,k}$, the sequence of event-triggering instants is defined iteratively by

$$k_{t+1}^i = \inf\{k \in \mathbb{Z}^+ | k > k_t^i, e_{i,k}^{\mathrm{T}} \Omega_{i,k}^{-1} e_{i,k} > 1\} \tag{9.10}$$

where $\Omega_{i,k} > 0$ $(i = 1, 2, \ldots, N)$ is referred to as the triggering threshold matrix of agent i at time instant k.

Remark 9.3 *The ellipsoidal triggering condition defined in (9.10) is quite general in that covers several well-studied triggering conditions as special cases. For example, it is observed from (9.10) that, when the matrix $\Omega_{i,k}$ is set to be a fixed positive scalar, then the ellipsoidal triggering condition specializes to the frequently used one as shown in [64] and the references therein. In particular, in the case that $\Omega_{i,k} \to 0$ (i.e. the size of ellipsoid approaches 0), the event-triggering mechanism will reduce to the traditional time-driven one. Moreover, another advantage of the proposed ellipsoidal triggering condition lies in the fact that the triggering threshold matrix $\Omega_{i,k}$ is actually a parameter that can be co-designed with the filter parameters, and this provides much flexibility in making trade-offs between the filtering performance and the triggering frequency, thereby achieving the balance between desired filtering accuracy and affordable resource consumption.*

By incorporating (9.9)-(9.10) with (9.7), we come up with the following *event-based* filter structure to be adopted in this chapter:

$$\hat{x}_{i,k+1} = F_{i,k}\hat{x}_{i,k} + \sum_{j \in \mathcal{N}_i} \theta_{ij} H_{ij,k} r_{j,k_t^j}, \ k \in [k_t^i, k_{t+1}^i) \tag{9.11}$$

where $F_{i,k}$ and $H_{ij,k}$ $(i, j \in \mathcal{V})$ are the filter parameters to be designed.

By (9.9)–(9.11), we have established the structure of a distributed filter with an event-triggering mechanism, which invokes the transmission of information when the difference between the current value and its latest transmitted value exceeds a certain threshold. Before proceeding further, we give the following assumption.

Assumption 9.3 *The initial state x_0 and its estimate $\hat{x}_{i,0}$ satisfy*

$$(x_0 - \hat{x}_{i,0})^\mathrm{T} P_0^{-1} (x_0 - \hat{x}_{i,0}) \leq 1 \tag{9.12}$$

where $P_0 > 0$ is a given positive definite matrix.

The objective of this chapter is twofold. *First*, for system (9.1) and filter (9.11), let the directed communication graph $\mathscr{G}$, the sequence of positive definite threshold matrices $\{\Omega_{i,k}\}_{k \geq 0}$ and the sequence of positive definite matrices $\{P_k\}_{k \geq 0}$ (constraints imposed on the filtering performance) be given. It is our first aim to design the sequences of filtering gains $\{F_{i,k}\}_{k \geq 0}$ and $\{H_{ij,k}\}_{k \geq 0}$ subject to the given triple $(\mathscr{G}, \{\Omega_{i,k}\}, \{P_k\})$ such that the following inequality is satisfied:

$$\Delta_{i,k} \triangleq (x_k - \hat{x}_{i,k})^\mathrm{T} P_k^{-1} (x_k - \hat{x}_{i,k}) \leq 1, \ i \in \mathscr{V}, \ k \geq 0. \tag{9.13}$$

Second, two optimization problems will be investigated for minimizing P_k and maximizing $\Omega_{i,k}$ in the sense of matrix trace at each time instant, respectively. This problem is referred to as a distributed event-based set-membership filtering problem.

9.2 Distributed Event-Based Set-Membership Filter Design

In this section, we will design a distributed event-based filter of form (9.11) for system (9.1) subject to sector-bounded nonlinearity, unknown but bounded noises and sensor saturations. A sufficient condition for the existence of the desired filter will be formulated in terms of a set of recursive linear matrix inequalities (RLMIs). First of all, we recall two useful lemmas for our following development.

Lemma 9.1 *(S-procedure [18]) Let $\psi_0(\cdot), \psi_1(\cdot), \ldots, \psi_p(\cdot)$ be quadratic functions of the variable $\varsigma \in \mathbb{R}^n$: $\psi_j(\varsigma) \triangleq \varsigma^\mathrm{T} X_j \varsigma$ $(j = 0, \ldots, p)$, where $X_j^\mathrm{T} = X_j$. If there exist $\epsilon_1 \geq 0, \ldots, \epsilon_p \geq 0$ such that $X_0 - \sum_{j=1}^{p} \epsilon_j X_j \leq 0$, then the following is true:*

$$\psi_1(\varsigma) \leq 0, \ldots, \psi_p(\varsigma) \leq 0 \rightarrow \psi_0(\varsigma) \leq 0. \tag{9.14}$$

Lemma 9.2 *(Schur Complement Equivalence) Given constant matrices* $\mathcal{S}_1, \mathcal{S}_2, \mathcal{S}_3$ *where* $\mathcal{S}_1 = \mathcal{S}_1^{\mathrm{T}}$ *and* $0 < \mathcal{S}_2 = \mathcal{S}_2^{\mathrm{T}}$, *then* $\mathcal{S}_1 + \mathcal{S}_3^{\mathrm{T}} \mathcal{S}_2^{-1} \mathcal{S}_3 < 0$ *if and only if*

$$\begin{bmatrix} \mathcal{S}_1 & \mathcal{S}_3^{\mathrm{T}} \\ \mathcal{S}_3 & -\mathcal{S}_2 \end{bmatrix} < 0 \quad \text{or} \quad \begin{bmatrix} -\mathcal{S}_2 & \mathcal{S}_3 \\ \mathcal{S}_3^{\mathrm{T}} & \mathcal{S}_1 \end{bmatrix} < 0. \tag{9.15}$$

Filter Design Subject to Fixed Triple $\left(\mathscr{G}, \{\Omega_{i,k}\}, \{P_k\}\right)$

For simplicity of notation, before giving the main results, we denote

$$\xi_k \triangleq \mathrm{col}_N\{x_k\},$$
$$\hat{x}_k \triangleq \mathrm{col}_N\{\hat{x}_{i,k}\},$$
$$e_k \triangleq \mathrm{col}_N\{e_{i,k}\},$$
$$\tilde{f}_k \triangleq \mathrm{col}_N\{f(x_k)\},$$
$$\tilde{\varphi}_k \triangleq \mathrm{col}_N\{\varphi(C_{i,k}x_k)\},$$
$$\mathcal{G}_k \triangleq \mathrm{diag}_N\{G_{1i}C_{i,k}\},$$
$$\mathcal{C}_k \triangleq \mathrm{diag}_N\{C_{i,k}\},$$
$$\mathcal{E}_k \triangleq \mathrm{diag}_N\{E_{i,k}\},$$
$$\mathcal{F}_k \triangleq \mathrm{diag}_N\{F_{i,k}\},$$
$$\Phi_{\varrho,i} \triangleq \mathrm{diag}\{\underbrace{0_\varrho, \ldots, 0_\varrho}_{i-1}, I_\varrho, \underbrace{0_\varrho, \ldots, 0_\varrho}_{N-i}\},$$
$$\mathcal{L}_{\varrho,i} \triangleq (1_N^{\mathrm{T}} \otimes I_\varrho)\Phi_{\varrho,i}, \quad \varrho = \{n, q, m\}.$$

From system (9.1) and filter (9.11), the one-step-ahead estimation error is obtained as follows:

$$\begin{aligned}
&x_{k+1} - \hat{x}_{i,k+1} \\
&= A_k x_k + D_k w_k + f(x_k) \\
&\quad - \left(F_{i,k}\hat{x}_{i,k} + \sum_{j \in \mathcal{N}_i} \theta_{ij} H_{ij,k} r_{j,k_t^j} \right) \\
&= A_k x_k + D_k w_k + f(x_k) - F_{i,k}\hat{x}_{i,k} \\
&\quad - \Big(\sum_{j \in \mathcal{N}_i} \theta_{ij} H_{ij,k} \big(\sigma(C_{j,k}x_k) + E_{j,k}v_k \\
&\quad - C_{j,k}\hat{x}_{j,k} + e_{j,k} \big) \Big). \tag{9.16}
\end{aligned}$$

By denoting $\tilde{x}_{i,k} \triangleq x_k - \hat{x}_{i,k}$ and $\tilde{x}_k \triangleq \mathrm{col}_N\{\tilde{x}_{i,k}\}$, we rewrite the filtering error dynamics (9.16) into the following compact form:

$$\begin{aligned}
\tilde{x}_{k+1} =&(I_N \otimes A_k)\xi_k + (1_N \otimes D_k)w_k + \tilde{f}_k \\
&- \mathcal{F}_k\hat{x}_k - \mathcal{H}_k\mathcal{G}_k\xi_k - \mathcal{H}_k\tilde{\varphi}_k \\
&- \mathcal{H}_k\mathcal{E}_k(1_N \otimes I_\nu)v_k + \mathcal{H}_k\mathcal{C}_k\hat{x}_k - \mathcal{H}_ke_k
\end{aligned} \tag{9.17}$$

where $\mathcal{H}_k \triangleq \left[\theta_{ij}H_{ij,k}\right]_{N\times N}$. Obviously, since $\theta_{ij} = 0$ when $j \notin \mathcal{N}_i$, $\mathcal{H}_k$ is a sparse matrix which can be expressed as

$$\mathcal{H}_k \in \mathcal{T}_{n\times m} \tag{9.18}$$

where $\mathcal{T}_{n\times m} \triangleq \left\{\mathcal{T} = [T_{ij}] \in \mathbb{R}^{nN \times mN} \middle| T_{ij} \in \mathbb{R}^{n\times m}, T_{ij} = 0 \text{ if } j \notin \mathcal{N}_i\right\}$.

The following theorem gives a sufficient condition for the solvability of the addressed distributed event-based set-membership filtering problem.

Theorem 9.1 *For system (9.1) and filter (9.11), let the triple $\left(\mathcal{G}, \{\Omega_{i,k}\}, \{P_k\}\right)$ be given. The design objective (9.13) is achieved if there exist sequences of real-valued matrices $\{\mathcal{F}_k\}_{k\geq 0}$ and $\{\mathcal{H}_k\}_{k\geq 0}$ $(\mathcal{H}_k \in \mathcal{T}_{n\times m})$, sequences of non-negative scalars $\{\epsilon_{i,k}^{(1)}\}_{k\geq 0}, \{\epsilon_{i,k}^{(2)}\}_{k\geq 0}, \{\epsilon_k^{(3)}\}_{k\geq 0}, \{\epsilon_k^{(4)}\}_{k\geq 0}, \{\epsilon_{i,k}^{(5)}\}_{k\geq 0}$ and $\{\epsilon_{i,k}^{(6)}\}_{k\geq 0}$ $(i = 1, 2, \ldots, N)$ satisfying the following N recursive linear matrix inequalities:*

$$\begin{bmatrix} -\Gamma_k & \Pi_k^{\mathrm{T}}\mathcal{L}_{n,i}^{\mathrm{T}} \\ \mathcal{L}_{n,i}\Pi_k & -P_{k+1} \end{bmatrix} \leq 0 \tag{9.19}$$

where

$$\Gamma_k = \sum_{i=1}^{N}\left(\epsilon_{i,k}^{(5)}\Xi_{i,k} + \epsilon_{i,k}^{(6)}\Psi_{i,k}\right) + \bar{\Gamma}_k, \tag{9.20}$$

$$\bar{\Gamma}_k = \operatorname{diag}\Big\{1 - \sum_{i=1}^{N}\left(\epsilon_{i,k}^{(1)} + \epsilon_{i,k}^{(2)}\right) - \epsilon_k^{(3)} - \epsilon_k^{(4)},$$

$$\sum_{i=1}^{N}\epsilon_{i,k}^{(1)}\mathcal{L}_{q,i}^{\mathrm{T}}\mathcal{L}_{q,i}, \sum_{i=1}^{N}\epsilon_{i,k}^{(2)}\mathcal{L}_{m,i}^{\mathrm{T}}\Omega_{i,k}^{-1}\mathcal{L}_{m,i},$$

$$\epsilon_k^{(3)}S_k^{-1}, \epsilon_k^{(4)}R_k^{-1}, 0, 0\Big\},$$

$$\Xi_{i,k} = \begin{bmatrix} \hat{x}_{i,k}^{\mathrm{T}}U_1^{\mathrm{T}}U_2\hat{x}_{i,k} & \hat{x}_k^{\mathrm{T}}\left(I_N \otimes (\bar{U}Q_k)\right)\Phi_{q,i} \\ * & I_N \otimes (Q_k^{\mathrm{T}}\bar{U}Q_k)\Phi_{q,i} \\ * & * \\ * & * \\ * & * \\ * & * \\ * & * \end{bmatrix}$$

$$
\begin{bmatrix}
0 & 0 & 0 & -\hat{x}_k^{\mathrm{T}}(I_N \otimes \tilde{U})\Phi_{n,i} & 0 \\
0 & 0 & 0 & -I_N \otimes (Q_k^{\mathrm{T}}\tilde{U})\Phi_{n,i} & 0 \\
0 & 0 & 0 & 0 & 0 \\
* & 0 & 0 & 0 & 0 \\
* & * & 0 & 0 & 0 \\
* & * & * & I_{nN}\Phi_{n,0} & 0 \\
* & * & * & * & 0
\end{bmatrix}, (9.21)
$$

$$
\bar{U} = \frac{U_1^{\mathrm{T}}U_2 + U_2^{\mathrm{T}}U_1}{2}, \qquad \tilde{U} = \frac{U_1^{\mathrm{T}} + U_2^{\mathrm{T}}}{2},
$$

$$
\Psi_{i,k} = \frac{1}{2}\begin{bmatrix} 0 & 0 & 0 & 0 & 0 & 0 & \bar{\Psi}_{i,k} \end{bmatrix}, \qquad (9.22)
$$

$$
\bar{\Psi}_{i,k} = \begin{bmatrix}
-\hat{x}_k^{\mathrm{T}}\mathrm{diag}\{C_{i,k}^{\mathrm{T}}G_i^{\mathrm{T}}\}\Phi_{m,i} \\
-\mathrm{diag}\{Q_k^{\mathrm{T}}C_{i,k}^{\mathrm{T}}G_i^{\mathrm{T}}\}\Phi_{m,i} \\
0 \\
0 \\
0 \\
0 \\
2I_{mN}\Phi_{m,i}
\end{bmatrix},
$$

$$
\Pi_k = \begin{bmatrix} \Pi_{11} & \Pi_{12} & -\mathcal{H}_k & \Pi_{14} & \Pi_{15} & I_{nN} & -\mathcal{H}_k \end{bmatrix}, \qquad (9.23)
$$

$$
\Pi_{11} = (I_N \otimes A_k - \mathcal{F}_k - \mathcal{H}_k(\mathcal{G}_k - \mathcal{C}_k))\,\hat{x}_k,
$$

$$
\Pi_{12} = I_N \otimes (A_k Q_k) - \mathcal{H}_k\mathcal{G}_k(I_N \otimes Q_k),
$$

$$
\Pi_{14} = \mathbf{1}_N \otimes D_k, \qquad \Pi_{15} = -\mathcal{H}_k\mathcal{E}_k(\mathbf{1}_N \otimes I_\nu),
$$

with $Q_k \in \mathbb{R}^{n \times q}$ being a factorization of P_k $(i.e., P_k = Q_k Q_k^{\mathrm{T}})$.

Proof *The proof is performed by induction. First, it can be immediately known from Assumption 9.3 that $\Delta_{i,0} \leq 1$ holds. Then, suppose that $\Delta_{i,k} \leq 1$ is true at time instant $k > 0$; we shall proceed to prove that $\Delta_{i,k+1} \leq 1$ holds.*

According to [66], since $\Delta_{i,k} \leq 1$ and $P_k = Q_k Q_k^{\mathrm{T}}$, there exist $z_{i,k} \in \mathbb{R}^q$ $(i = 1, 2, \ldots, N)$ with $\|z_{i,k}\| \leq 1$ such that

$$
x_k = \hat{x}_{i,k} + Q_k z_{i,k}. \qquad (9.24)
$$

Obviously, by denoting $z_k \triangleq \mathrm{col}_N\{z_{i,k}\}$, (9.24) is equivalent to

$$
\xi_k = \hat{x}_k + (I_N \otimes Q_k)z_k. \qquad (9.25)
$$

From (9.24) and (9.25), the filtering error system (9.17) is rewritten as follows:

$$
\begin{aligned}
\tilde{x}_{k+1} =& (I_N \otimes A_k)\big(\hat{x}_k + (I_N \otimes Q_k)z_k\big) \\
& + (\mathbf{1}_N \otimes D_k)w_k + \tilde{f}_k - \mathcal{F}_k\hat{x}_k \\
& - \mathcal{H}_k\mathcal{G}_k\big(\hat{x}_k + (I_N \otimes Q_k)z_k\big) \\
& - \mathcal{H}_k\tilde{\varphi}_k - \mathcal{H}_k\mathcal{E}_k(\mathbf{1}_N \otimes I_\nu)v_k
\end{aligned}
$$

$$+ \mathcal{H}_k \mathcal{C}_k \hat{x}_k - \mathcal{H}_k e_k$$

$$= (I_N \otimes A_k - \mathcal{F}_k - \mathcal{H}_k(\mathcal{G}_k - \mathcal{C}_k)) \hat{x}_k$$

$$+ (I_N \otimes (A_k Q_k) - \mathcal{H}_k \mathcal{G}_k (I_N \otimes Q_k)) z_k$$

$$+ (\mathbf{1}_N \otimes D_k) w_k + \tilde{f}_k - \mathcal{H}_k \tilde{\varphi}_k$$

$$- \mathcal{H}_k \mathcal{E}_k (\mathbf{1}_N \otimes I_\nu) v_k - \mathcal{H}_k e_k. \tag{9.26}$$

Denoting

$$\eta_k \triangleq \begin{bmatrix} 1 \\ z_k \\ e_k \\ w_k \\ v_k \\ \tilde{f}_k \\ \tilde{\varphi}_k \end{bmatrix}, \tag{9.27}$$

the filtering error dynamics can be further expressed by

$$\tilde{x}_{k+1} = \Pi_k \eta_k \tag{9.28}$$

where Π_k is defined in (9.23).

It follows from (9.2), (9.10) and (9.24) that the vectors $z_{i,k}$, $e_{i,k}$, w_k and v_k are satisfying

$$\begin{cases} \|z_{i,k}\| \leq 1 \\ e_{i,k}^{\mathrm{T}} \Omega_{i,k}^{-1} e_{i,k} \leq 1 \\ w_k^{\mathrm{T}} S_k^{-1} w_k \leq 1 \\ v_k^{\mathrm{T}} R_k^{-1} v_k \leq 1 \end{cases} \tag{9.29}$$

which, by (9.27), can be rewritten in terms of η_k as follows:

$$\begin{cases} \eta_k^{\mathrm{T}} \mathrm{diag} \left\{ -1, \mathcal{L}_{q,i}^{\mathrm{T}} \mathcal{L}_{q,i}, 0, 0, 0, 0, 0 \right\} \eta_k \leq 0 \\ \eta_k^{\mathrm{T}} \mathrm{diag} \left\{ -1, 0, \mathcal{L}_{m,i}^{\mathrm{T}} \Omega_{i,k}^{-1} \mathcal{L}_{m,i}, 0, 0, 0, 0 \right\} \eta_k \leq 0 \\ \eta_k^{\mathrm{T}} \mathrm{diag} \left\{ -1, 0, 0, S_k^{-1}, 0, 0, 0 \right\} \eta_k \leq 0 \\ \eta_k^{\mathrm{T}} \mathrm{diag} \left\{ -1, 0, 0, 0, R_k^{-1}, 0, 0 \right\} \eta_k \leq 0 \ . \end{cases} \tag{9.30}$$

We now proceed to investigate the sector-bounded nonlinearity $f(x_k)$ in system (9.1). From Assumption 9.2, $f(x_k)$ belongs to sector $[U_1, U_2]$, which can be formulated by

$$\left(f(x_k) - U_1 x_k\right)^{\mathrm{T}} \left(f(x_k) - U_2 x_k\right) \leq 0. \tag{9.31}$$

Substituting (9.24) into (9.31) results in

$$f^{\mathrm{T}}(x_k)f(x_k) + \hat{x}_{i,k}^{\mathrm{T}}U_1^{\mathrm{T}}U_2\hat{x}_{i,k} + z_{i,k}^{\mathrm{T}}Q_k^{\mathrm{T}}U_1^{\mathrm{T}}U_2Q_kz_{i,k}$$
$$+\hat{x}_{i,k}^{\mathrm{T}}U_1^{\mathrm{T}}U_2Q_kz_{i,k} + z_{i,k}^{\mathrm{T}}Q_k^{\mathrm{T}}U_1^{\mathrm{T}}U_2\hat{x}_{i,k} - f^{\mathrm{T}}(x_k)U_2\hat{x}_{i,k}$$
$$-f^{\mathrm{T}}(x_k)U_2Q_kz_{i,k} - \hat{x}_{i,k}^{\mathrm{T}}U_1^{\mathrm{T}}f(x_k)$$
$$-z_{i,k}^{\mathrm{T}}Q_k^{\mathrm{T}}U_1^{\mathrm{T}}f(x_k) \le 0 \tag{9.32}$$

which can be expressed by η_k as

$$\eta_k^{\mathrm{T}}\Xi_{i,k}\eta_k \le 0 \tag{9.33}$$

with $\Xi_{i,k}$ being defined in (9.21).

By the same token, we have from the sensor saturation constraints in (9.6) that

$$\varphi^{\mathrm{T}}(C_{i,k}x_k)\big(\varphi(C_{i,k}x_k) - G_iC_{i,k}(\hat{x}_{i,k} + Q_kz_{i,k})\big) \le 0 \tag{9.34}$$

which can be formulated by η_k as

$$\eta_k^{\mathrm{T}}\Psi_{i,k}\eta_k \le 0 \tag{9.35}$$

with $\Psi_{i,k}$ being defined in (9.22).

In the following, according to the principle of induction, it remains to show that $\Delta_{i,k+1} \le 1$ is true if the condition of this theorem is satisfied at time instant k; i.e., there exist real-valued matrices $\mathcal{F}_k$ and $\mathcal{H}_k$ $(\mathcal{H}_k \in \mathscr{T}_{n\times m})$, non-negative scalars $\epsilon_{i,k}^{(1)} \ge 0$, $\epsilon_{i,k}^{(2)} \ge 0$, $\epsilon_k^{(3)} \ge 0$, $\epsilon_k^{(4)} \ge 0$, $\epsilon_{i,k}^{(5)} \ge 0$ and $\epsilon_{i,k}^{(6)} \ge 0$ $(i = 1, 2, \ldots, N)$ satisfying RLMIs (9.19).

By resorting to the Schur Complement Equivalence (Lemma 9.2), it can be seen that the set of RLMIs (9.19) holds if and only if

$$\Pi_k^{\mathrm{T}}\mathcal{L}_{n,i}^{\mathrm{T}}P_{k+1}^{-1}\mathcal{L}_{n,i}\Pi_k - \Gamma_k \le 0 \tag{9.36}$$

which, by (9.20), is equivalent to

$$\Pi_k^{\mathrm{T}}\mathcal{L}_{n,i}^{\mathrm{T}}P_{k+1}^{-1}\mathcal{L}_{n,i}\Pi_k - \mathrm{diag}\{1,0,0,0,0,0,0\}$$
$$-\sum_{i=1}^{N}\epsilon_{i,k}^{(1)}\mathrm{diag}\{-1, \mathcal{L}_{q,i}^{\mathrm{T}}\mathcal{L}_{q,i}, 0,0,0,0,0\}$$
$$-\sum_{i=1}^{N}\epsilon_{i,k}^{(2)}\mathrm{diag}\left\{-1, 0, \mathcal{L}_{m,i}^{\mathrm{T}}\Omega_{i,k}^{-1}\mathcal{L}_{m,i}, 0,0,0,0\right\}$$
$$-\epsilon_k^{(3)}\mathrm{diag}\left\{-1,0,0,S_k^{-1},0,0,0\right\}$$
$$-\epsilon_k^{(4)}\mathrm{diag}\left\{-1,0,0,0,R_k^{-1},0,0\right\}$$
$$-\sum_{i=1}^{N}\epsilon_{i,k}^{(5)}\Xi_{i,k} - \sum_{i=1}^{N}\epsilon_{i,k}^{(6)}\Psi_{i,k} \le 0. \tag{9.37}$$

In view of (9.30), (9.33) and (9.35), it follows directly from the S-procedure (Lemma 9.1) that

$$\eta_k^{\mathrm{T}}\left(\Pi_k^{\mathrm{T}}\mathcal{L}_{n,i}^{\mathrm{T}}P_{k+1}^{-1}\mathcal{L}_{n,i}\Pi_k\right.$$
$$\left. - \operatorname{diag}\{1,0,0,0,0,0,0\}\right)\eta_k \leq 0. \tag{9.38}$$

It is evident that the following equivalences hold:

$$\text{Inequality } (9.38)$$
$$\Longleftrightarrow \eta_k^{\mathrm{T}}\Pi_k^{\mathrm{T}}\mathcal{L}_{n,i}^{\mathrm{T}}P_{k+1}^{-1}\mathcal{L}_{n,i}\Pi_k\eta_k \leq 1$$
$$\overset{(9.28)}{\Longleftrightarrow} \tilde{x}_{k+1}^{\mathrm{T}}\mathcal{L}_{n,i}^{\mathrm{T}}P_{k+1}^{-1}\mathcal{L}_{n,i}\tilde{x}_{k+1} \leq 1$$
$$\Longleftrightarrow (x_{k+1} - \hat{x}_{i,k+1})^{\mathrm{T}}P_{k+1}^{-1}(x_{k+1} - \hat{x}_{i,k+1}) \leq 1. \tag{9.39}$$

We can now conclude from (9.39) that $\Delta_{i,k+1} \leq 1$ is achieved, and the induction is accomplished. Therefore, the design objective (9.13) is met with the obtained sequences of parameters $\{F_k\}_{k\geq 0}$ and $\{\mathcal{H}_k\}_{k\geq 0}$ for fixed triple $(\mathscr{G}, \{\Omega_{i,k}\}, \{P_k\})$. The proof is now complete.

In the following, an iterative algorithm is presented to compute the sequences of the filtering parameters $\{F_{i,k}\}_{k\geq 0}$ and $\{H_{ij,k}\}_{k\geq 0}$ recursively.

Algorithm 9.1 *Computational Algorithm for $\{F_{i,k}\}_{k\geq 0}$ and $\{H_{ij,k}\}_{k\geq 0}$*

1) *Initialization: Set $k = 0$ and the maximum computation step $k_{\max}$. Set the triple $(\mathscr{G}, \{\Omega_{i,k}\}, \{P_k\})$ for $0 \leq k \leq k_{\max}$. Then factorize $\{P_k\}$ appropriately to obtain the sequence of matrices $\{Q_k\}$. Select the initial values of x_0 and $\hat{x}_{i,0}$ satisfying (9.12). Then $\hat{x}_0 = \operatorname{col}_N\{\hat{x}_{i,0}\}$ is known.*

2) *With the obtained $\hat{x}_k$ and Q_k, solve the RLMIs (9.19) for F_k and $\mathcal{H}_k$. Then $F_{i,k}$ and $H_{ij,k}$ can be obtained.*

3) *With the obtained F_k and $\mathcal{H}_k$, compute $\hat{x}_{i,k+1}$ according to (9.11). Then $\hat{x}_{k+1} = \operatorname{col}_N\{\hat{x}_{i,k+1}\}$ is obtained.*

4) *Set $k = k + 1$. If $k > k_{\max}$, exit. Otherwise, go to 2).*

Filter Design Subject to Constraint of Average Filtering Errors

In many cases, from a global point of view, we are more interested in the filtering performance in terms of an average of estimation errors among all the sensing nodes rather than the individual ones; see [193, 207] for references. As such, in this subsection, based on the results obtained so far, we will further discuss the distributed filtering problem subject to constraints imposed on the average of filtering errors. To begin with, we define the average filtering error ζ_k as follows:

$$\zeta_k \triangleq \sum_{i=1}^{N} \lambda_i (x_k - \hat{x}_{i,k})$$
$$= (\mathbf{1}^{\mathrm{T}} \Lambda \otimes I_n)(\xi_k - \hat{x}_k)$$
$$= (\mathbf{1}^{\mathrm{T}} \Lambda \otimes I_n)\tilde{x}_k \tag{9.40}$$

where the weighting parameters λ_i $(i = 1, 2, \ldots, N)$ represent the priorities with respect to the corresponding sensing nodes.

Assume $\zeta_0^{\mathrm{T}} Y_0^{-1} \zeta_0 \leq 1$ where $Y_0 > 0$ is a given matrix. Let the triple $(\mathscr{G}, \{\Omega_{i,k}\}, \{Y_k\})$ be given, where $\{Y_k\}_{k\geq 0}$ is a sequence of positive definite matrices describing the constraints imposed on the average filtering performance. It is our objective in this subsection to design the filter parameters in (9.11) such that the following requirement is met for $k \geq 0$:

$$\zeta_k^{\mathrm{T}} Y_k^{-1} \zeta_k \leq 1. \tag{9.41}$$

Theorem 9.2 *For system* (9.1) *and filter* (9.11), *let the triple* $(\mathscr{G}, \{\Omega_{i,k}\}, \{Y_k\})$ *and the initial condition* $\zeta_0^{\mathrm{T}} Y_0^{-1} \zeta_0 \leq 1$ *be given. The requirement* (9.41) *is achieved if there exist sequences of real-valued matrices* $\{\mathcal{F}_k\}_{k\geq 0}$ *and* $\{\mathcal{H}_k\}_{k\geq 0}$ $(\mathcal{H}_k \in \mathcal{T}_{n\times m})$, *sequences of non-negative scalars* $\{\epsilon_{i,k}^{(1)}\}_{k\geq 0}$, $\{\epsilon_{i,k}^{(2)}\}_{k\geq 0}$, $\{\epsilon_k^{(3)}\}_{k\geq 0}$, $\{\epsilon_k^{(4)}\}_{k\geq 0}$, $\{\epsilon_{i,k}^{(5)}\}_{k\geq 0}$ *and* $\{\epsilon_{i,k}^{(6)}\}_{k\geq 0}$ $(i = 1, 2, \ldots, N)$ *satisfying the following RLMI:*

$$\begin{bmatrix} -\Gamma_k & \Pi_k^{\mathrm{T}}(\mathbf{1}^{\mathrm{T}}\Lambda \otimes I_n)^{\mathrm{T}} \\ (\mathbf{1}^{\mathrm{T}}\Lambda \otimes I_n)\Pi_k & -Y_{k+1} \end{bmatrix} \leq 0 \tag{9.42}$$

where Γ_k *and* Π_k *are defined in* (9.20) *and* (9.23), *respectively.*

Proof *With the fixed triple* $(\mathscr{G}, \{\Omega_{i,k}\}, \{Y_k\})$ *and given the initial condition* $\zeta_0^{\mathrm{T}} Y_0^{-1} \zeta_0 \leq 1$, *the proof of Theorem 9.2 can be accomplished by induction which is analogous to that of Theorem 9.1 and is therefore omitted here.*

Optimization Algorithms

Theorems 9.1 and 9.2 in previous subsections outline the principles of designing the filtering parameters by solving the corresponding set of RLMIs. It should be pointed out, however, that neither of the proposed methodologies provides an optimal solution. As discussed previously, we now proceed to deal with the second part of our design objective, that is, minimizing $\{P_k\}_{k\geq 0}$ (for the locally best filtering performance) and maximizing $\{\Omega_{i,k}\}_{k\geq 0}$ (for the locally lowest triggering frequency) in the sense of matrix trace, respectively. In the following stage, two optimization problems based on Theorem 9.1 will be proposed to demonstrate the flexibility of our developed strategy. Such a kind of flexibility allows us to make a compromise between filtering performance and triggering frequency to achieve a balance between accuracy and cost.

Optimization Problem 1: *Minimization of* $\{P_k\}_{k\geq0}$ *(in the sense of matrix trace) for the locally best filtering performance.*

Corollary 9.1 *For the discrete time-varying nonlinear system (9.1) with filter (9.11), let the pair* $\left(\mathscr{G}, \{\Omega_{i,k}\}\right)$ *be given. A sequence of minimized* $\{P_k\}_{k\geq0}$ *(in the sense of matrix trace) is guaranteed if there exist sequences of real-valued matrices* $\{\mathcal{F}_k\}_{k\geq0}$ *and* $\{\mathcal{H}_k\}_{k\geq0}$ $\left(\mathcal{H}_k \subset \mathscr{T}_{n\times m}\right)$, *sequences of non-negative scalars* $\{\epsilon_{i,k}^{(1)}\}_{k\geq0}, \{\epsilon_{i,k}^{(2)}\}_{k\geq0}, \{\epsilon_k^{(3)}\}_{k\geq0}, \{\epsilon_k^{(4)}\}_{k\geq0}, \{\epsilon_{i,k}^{(5)}\}_{k\geq0}$ *and* $\{\epsilon_{i,k}^{(6)}\}_{k\geq0}$ $(i = 1, 2, \ldots, N)$ *solving the following optimization problem:*

$$\min_{P_{k+1},\mathcal{F}_k,\mathcal{H}_k,\epsilon_{i,k}^{(1)},\epsilon_{i,k}^{(2)},\epsilon_k^{(3)},\epsilon_k^{(4)},\epsilon_{i,k}^{(5)},\epsilon_{i,k}^{(6)}} \operatorname{tr}[P_{k+1}] \tag{9.43}$$

subject to

$$\begin{bmatrix} -\Gamma_k & \Pi_k^{\mathrm{T}}\mathcal{L}_{n,i}^{\mathrm{T}} \\ \mathcal{L}_{n,i}\Pi_k & -P_{k+1} \end{bmatrix} \leq 0. \tag{9.44}$$

Notice that the inequalities (9.44) are linear to the variables $P_{k+1}, \mathcal{F}_k, \mathcal{H}_k,$ $\epsilon_{i,k}^{(1)}, \epsilon_{i,k}^{(2)}, \epsilon_k^{(3)}, \epsilon_k^{(4)}, \epsilon_{i,k}^{(5)}$ and $\epsilon_{i,k}^{(6)}$. Therefore, it follows directly from Corollary 9.1 that Optimization Problem 1 can be readily solved via the existing semidefinite programming methods [161].

Optimization Problem 2: *Maximization of triggering threshold matrices* $\{\Omega_{i,k}\}_{k\geq0}$ *(in the sense of matrix trace) for the locally lowest triggering frequency.*

Corollary 9.2 *For the discrete time-varying nonlinear system (9.1) with filter (9.11), let the pair* $\left(\mathscr{G}, \{P_k\}\right)$ *be given. The locally lowest triggering frequency can be determined if there exist sequences of positive definite matrices* $\{\Upsilon_{i,k}\}_{k\geq0}$, *sequences of real-valued matrices* $\{\mathcal{F}_k\}_{k\geq0}$ *and* $\{\mathcal{H}_k\}_{k\geq0}$ $\left(\mathcal{H}_k \in \mathscr{T}_{n\times m}\right)$, *non-negative scalars* $\{\epsilon_{i,k}^{(1)}\}_{k\geq0}, \{\epsilon_{i,k}^{(2)}\}_{k\geq0}, \{\epsilon_k^{(3)}\}_{k\geq0},$ $\{\epsilon_k^{(4)}\}_{k\geq0}, \{\epsilon_{i,k}^{(5)}\}_{k\geq0}$ *and* $\{\epsilon_{i,k}^{(6)}\}_{k\geq0}$ $(i = 1, 2, \ldots, N)$ *solving the following optimization problem:*

$$\min_{\mathcal{F}_k,\mathcal{H}_k,\Upsilon_{i,k},\epsilon_{i,k}^{(1)},\epsilon_{i,k}^{(2)},\epsilon_k^{(3)},\epsilon_k^{(4)},\epsilon_{i,k}^{(5)},\epsilon_{i,k}^{(6)}} \operatorname{tr}\left[\sum_{i=1}^{N} \beta_i \Upsilon_{i,k}\right] \tag{9.45}$$

subject to

$$\begin{bmatrix} -\tilde{\Gamma}_k & \Pi_k^{\mathrm{T}}\mathcal{L}_{n,i}^{\mathrm{T}} \\ \mathcal{L}_{n,i}\Pi_k & -P_{k+1} \end{bmatrix} \leq 0 \tag{9.46}$$

where

$$\tilde{\Gamma}_k = \sum_{i=1}^{N} \left(\epsilon_{i,k}^{(5)} \Xi_{i,k} + \epsilon_{i,k}^{(6)} \Psi_{i,k} \right)$$

$$+ \operatorname{diag}\left\{ 1 - \sum_{i=1}^{N} \left(\epsilon_{i,k}^{(1)} + \epsilon_{i,k}^{(2)} \right) - \epsilon_k^{(3)} - \epsilon_k^{(4)}, \right.$$

$$\sum_{i=1}^{N} \epsilon_{i,k}^{(1)} \mathcal{L}_{q,i}^{\mathrm{T}} \mathcal{L}_{q,i}, \sum_{i=1}^{N} \mathcal{L}_{m,i}^{\mathrm{T}} \epsilon_{i,k}^{(2)} \Upsilon_{i,k} \mathcal{L}_{m,i},$$

$$\left. \epsilon_k^{(3)} S_k^{-1}, \epsilon_k^{(4)} R_k^{-1}, 0, 0 \right\}$$

and $\beta_i > 0$ $(i = 1, 2, \dots, N)$ *are the weighting scalars satisfying* $\sum_{i=1}^{N} \beta_i = 1$. *The threshold matrix* $\Omega_{i,k}$ *at each time instant can be determined by* $\Omega_{i,k} = \Upsilon_{i,k}^{-1}$.

Remark 9.4 *With the satisfaction of a certain predetermined filtering performance (i.e., a prescribed sequence of* $\{P_k\}_{k \geq 0}$*), Corollary 9.2 presents a way to maximize a certain combination of individual triggering thresholds so as to reduce the triggering frequency. The values of weighting scalars* β_i $(i = 1, 2, \dots, N)$ *are usually set to be equal but they can be adjusted according to the priorities of certain sensing nodes. Similarly, by borrowing an idea from the proposed optimization problem (9.45)–(9.46), we are able to determine the "largest" noises that can be tolerated with the satisfaction of a certain prespecified filtering performance.*

Noting that (9.46) is actually a set of bilinear matrix inequalities (BMIs) because of the term $\epsilon_{i,k}^{(2)} \Upsilon_{i,k}$, in the following stage, we provide a numerical algorithm based on chaos optimization [77] to solve Optimization Problem 2. To start with, we introduce the following iterative chaotic mapping [77]:

$$\rho(\tau + 1) = \sin\left(\frac{\alpha}{\rho(\tau)} \right) \tag{9.47}$$

where $\rho(\tau) \in [-1, 0) \cup (0, 1]$ is a chaotic variable and $\alpha > 0$ is a properly selected parameter and τ $(\tau = 0, 1, 2, \dots)$ indicates the iterative counter.

Remark 9.5 *Recently, chaos optimization techniques have been employed to solve a variety of global optimization problems due to the properties of unique ergodicity and irregularity of the series generated by chaos [131]. Noticing that in the chaotic mapping (9.47), if the initial value is set as* $\rho(0) \neq 0$*, then the chaotic variable* $\rho(\tau)$ *is able to traverse every state in the interval* $[-1, 0) \cup (0, 1]$ *in an ergodic way, and each state is visited only once. As such, it makes* $\rho(\tau)$ *a potential candidate for solving the set of BMIs (9.46) iteratively.*

We now present a detailed algorithm to solve Optimization Problem 2. Let $\epsilon_{i,k}^{(2)}$ in (9.46) be chaotic variable. We first need to determine the interval over which $\epsilon_{i,k}^{(2)}$ can traverse ergodically. It can be easily seen that if (9.46) holds, then it follows directly that

$$-\tilde{\Gamma}_k \leq 0 \tag{9.48}$$

which indicates by (9.21) that

$$\bar{f}_k + 1 - \sum_{i=1}^{N} \left(\epsilon_{i,k}^{(1)} + \epsilon_{i,k}^{(2)} \right) - \epsilon_k^{(3)} - \epsilon_k^{(4)} \geq 0 \tag{9.49}$$

where $\bar{f}_k \triangleq \sum_{i=1}^{N} \epsilon_{i,k}^{f} \hat{x}_{i,k}^{\mathrm{T}} U_1^{\mathrm{T}} U_2 \hat{x}_{i,k}$.

Taking into account that $\epsilon_{i,k}^{(1)} \geq 0$, $\epsilon_{i,k}^{(2)} \geq 0$, $\epsilon_k^{(3)} \geq 0$ and $\epsilon_k^{(4)} \geq 0$, we acquire $0 \leq \epsilon_{i,k}^{(2)} \leq \bar{f}_k + 1$. In order to determine the maximum of $\bar{f}_k$, we propose the following auxiliary optimization problem:

$$\max_{P_{k+1},\mathcal{F}_k,\mathcal{H}_k,\epsilon_{i,k}^{(1)},\epsilon_{i,k}^{(2)},\epsilon_k^{(3)},\epsilon_k^{(4)},\epsilon_{i,k}^{(5)},\epsilon_{i,k}^{(6)}} \bar{f}_k \tag{9.50}$$

$$\text{subject to} \quad (9.44)$$

We present the optimization problem (9.50) to seek the maximum of $\bar{f}_k$ thereby determining the interval in which $\epsilon_{i,k}^{(2)}$ is confined. If it is solvable, then we denote the optimal value as $\bar{f}_k^*$, and it follows directly that $\epsilon_{i,k}^{(2)} \in (0, \bar{f}_k^*+1]$. Denoting the chaotic variable during the τth iteration as $\epsilon_{i,k}^{(2)}(\tau)$, and by taking (9.47) into consideration, it can be set as

$$\epsilon_{i,k}^{(2)}(\tau) = \frac{\bar{f}_k^* + 1}{2} \left(1 + \rho_i(\tau) \right), \qquad i = 1, 2, \ldots, N \tag{9.51}$$

With $\epsilon_{i,k}^{(2)}(\tau)$ obtained during the τth iteration, the addressed optimization problem (9.45) subject to constraint (9.46) is converted into a semi-definite programming problem with LMI constraints which can be effectively solved by existing tools.

For time instant k, denote $\phi_k\big(\epsilon_{i,k}^{(2)}\big) \triangleq \mathrm{tr}\left[\sum_{i=1}^{N} \beta_i \Upsilon_{i,k}\right]$. Then, during the τth iteration, the intermediate index $\phi_k\big(\epsilon_{i,k}^{(2)}(\tau)\big)$ is given as follows:

$$\phi_k\big(\epsilon_{i,k}^{(2)}(\tau)\big) = \begin{cases} \mathrm{tr}\left[\sum_{i=1}^{N} \beta_i \Upsilon_{i,k}(\tau)\right], & \text{if (9.45) is solvable} \\ \vartheta, & \text{otherwise} \end{cases}$$

where $\vartheta > 0$ is a sufficiently large constant. In the following, we will formulate the detailed algorithm for solving Optimization Problem 2.

Algorithm 9.2 *Algorithm for Solving Optimization Problem 2*

1) *Initialization: Set $k = 0$. Set the maximum step $k_{\max}$.*

2) *Set $\tau = 0$, $\phi_k^* = \vartheta$. Set the maximum iteration times $\tau_{\max}$ for the chaotic optimization and select the proper initial value of $\rho_i(0) \neq 0$.*

3) *Solve the optimization problem (9.50) and obtain $\bar{f}_k^*$.*

4) *Obtain $\epsilon_{i,k}^{(2)}(\tau)$ from equation (9.51) with known $\rho_i(\tau)$.*

5) *Solve the semi-definite programming problem (9.45) with constraint (9.46) by using the obtained $\epsilon_{i,k}^{(2)}(\tau)$. If $\phi_k(\epsilon_{i,k}^{(2)}(\tau)) < \phi_k^*$, then let $\phi_k^* = \phi_k(\epsilon_{i,k}^{(2)}(\tau))$ and $\epsilon_{i,k}^* = \epsilon_{i,k}^{(2)}(\tau)$. Otherwise, go to 6).*

6) *Set $\tau = \tau + 1$. Calculate $\rho_i(\tau)$ according to (9.47).*

7) *If $\tau > \tau_{\max}$, or ϕ_k^* does not change after certain iteration times, output the optimal values and go to 8). Otherwise, go to 4).*

8) *Set $k = k + 1$. If $k > k_{\max}$, exit. Otherwise, go to 3).*

Up to now, the addressed distributed event-based set-membership filtering problem has been discussed for nonlinear systems subject to unknown but bounded noises and sensor saturations. In terms of the feasibility of certain RLMIs, the sufficient conditions for the solvability of the addressed problem have been presented. Moreover, in order to show the advantage of our developed algorithm, two optimization problems have been investigated to demonstrate the flexibility of the filter design technique which is capable of making a compromise between filtering accuracy and communication cost. It will be further verified later in Section 9.3 by an illustrative example.

Remark 9.6 *The advantages of the developed method can be summarized as follows: i) within the established generic framework, the sector-bounded non-linearity, unknown but bounded noises, and sensor saturations can be tackled simultaneously with the proposed ellipsoidal triggering condition; ii) it allows much flexibility in making trade-offs between filtering accuracy and communication cost, while both of the essential objectives can be met at the same time; iii) the proposed method has the potential to deal with the distributed filtering problem over sensor networks with time-varying topology.*

9.3   An Illustrative Example

In this section, an illustrative example is presented to show the validity of the proposed filter design strategy. Consider a nonlinear discrete time-varying nonlinear system with 3 sensing nodes that has the following parameters:

$$A_k = \begin{bmatrix} 0.55 + 0.11\sin(0.5k) & 0.01 + 0.01\sin(2k) \\ 0.01 & 0.55 + 0.11\sin(0.5k) \end{bmatrix},$$

$$D_k = \begin{bmatrix} -0.1 + 0.05\cos(3k) \\ 0.2 + 0.04e^{-k} \end{bmatrix},$$

$$C_{1,k} = \begin{bmatrix} 0.73 + 0.2\sin(k) & 0.1 \end{bmatrix},$$

$$E_{1,k} = 0.2 + 0.05\cos(3k),$$

$$C_{2,k} = \begin{bmatrix} 0.1 & 0.75 + 0.4\sin(2k) \end{bmatrix},$$

$$E_{2,k} = 0.2 + 0.15\sin(2k),$$

$$C_{3,k} = \begin{bmatrix} 0.75 & 0.1 \end{bmatrix},$$

$$E_{3,k} = 0.15 + 0.05\sin(2k).$$

Suppose that the sensor network is represented by a directed graph $\mathscr{G} = (\mathscr{V}, \mathscr{E}, \mathscr{L})$ where the set of nodes $\mathscr{V} = \{1, 2, 3\}$, the set of edges $\mathscr{E} = \{(1,1), (1,2), (1,3), (2,1), (2,2), (2,3), (3,1), (3,2), (3,3)\}$, and the adjacency elements associated with the edges of the graph are $\theta_{ij} = 1$.

Let the disturbances be $w_k = 1.2\sin(2k)$ and $v_k = 1.5\cos(5k)$ and set $S = 2$ and $R = 3$. It then can be easily checked that w_k and v_k belong to the ellipsoidal sets defined in (9.2). Let the initial values be given as follows:

$$x_0 = \begin{bmatrix} 5 \\ 3 \end{bmatrix}, \quad \hat{x}_{1,0} = \hat{x}_{2,0} = \hat{x}_{3,0} = \begin{bmatrix} 5.02 \\ 3.01 \end{bmatrix},$$

$$P_0 = \begin{bmatrix} 0.1 & 0.01 \\ 0.01 & 0.1 \end{bmatrix}.$$

Then it can be easily verified that Assumption 9.3 is satisfied.

The nonlinear function $f(x_k)$ is chosen as

$$f(x_k) = \begin{bmatrix} -0.1x_k^{(1)} + 0.15x_k^{(2)} + \dfrac{0.1x_k^{(2)}\sin(x_k^{(1)})}{\sqrt{(x_k^{(1)})^2 + (x_k^{(2)})^2 + 10}} \\ -0.05x_k^{(1)} + 0.05x_k^{(2)} \end{bmatrix}$$

where $x_k^{(1)} \triangleq [1 \ 0]x_k$ and $x_k^{(2)} \triangleq [0 \ 1]x_k$ represent the first and second entries of the system state, respectively. It can be verified that $f(x_k)$ belongs to the sector $[U_1, U_2]$ with

$$U_1 = \begin{bmatrix} -0.4 & 0 \\ -0.2 & -0.3 \end{bmatrix},$$

$$U_2 = \begin{bmatrix} 0.2 & 0.3 \\ 0.1 & 0.4 \end{bmatrix}.$$

Denote the first and second entries of $C_{i,k}x_k$ as $\tilde{y}_{i,k}^{(1)}$ and $\tilde{y}_{i,k}^{(2)}$, respectively. With the purpose of showing the influence of sensor saturation on the filtering performance, we consider the following two cases:

Case 1:

$$\begin{cases} \sigma\big(\tilde{y}_{1,k}^{(1)}\big)_{\max} = \sigma\big(\tilde{y}_{1,k}^{(2)}\big)_{\max} = 20, \\ \sigma\big(\tilde{y}_{2,k}^{(1)}\big)_{\max} = \sigma\big(\tilde{y}_{2,k}^{(2)}\big)_{\max} = 30, \\ \sigma\big(\tilde{y}_{3,k}^{(1)}\big)_{\max} = \sigma\big(\tilde{y}_{3,k}^{(2)}\big)_{\max} = 20. \end{cases} \qquad (9.52)$$

Case 2:

$$\begin{cases} \sigma\big(\tilde{y}_{1,k}^{(1)}\big)_{\max} = \sigma\big(\tilde{y}_{1,k}^{(2)}\big)_{\max} = 10, \\ \sigma\big(\tilde{y}_{2,k}^{(1)}\big)_{\max} = \sigma\big(\tilde{y}_{2,k}^{(2)}\big)_{\max} = 15, \\ \sigma\big(\tilde{y}_{3,k}^{(1)}\big)_{\max} = \sigma\big(\tilde{y}_{3,k}^{(2)}\big)_{\max} = 10. \end{cases} \qquad (9.53)$$

In this section, we proceed to utilize the algorithms proposed in Corollary 9.1 and Corollary 9.2 to solve Optimization Problem 1 (*OP1*) and Optimization Problem 2 (*OP2*), respectively. The simulations are performed by means of Matlab software (YALMIP 3.0), and the results are shown in Figs. 9.1–9.8.

Fig. 9.1 and Fig. 9.2 depict the time instants of each agent when the trigger occurs in *OP1* and *OP2*, respectively. By comparison of triggering times shown in the figures, we can obviously see that 1) the proposed event-triggering mechanism can effectively reduce the frequency of innovation broadcasting; 2) the total triggering times in *OP2* are much less than that in *OP1*, indicating that the triggering frequency can be further reduced if we implement the strategy provided in *OP2*, exactly as anticipated. For the system subject to the saturation in *Case 1*, the trajectories of the estimation errors of the system

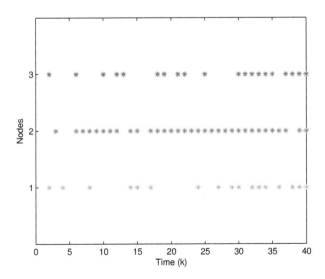

FIGURE 9.1: The triggering sequences for **OP1** (*Case 1*).

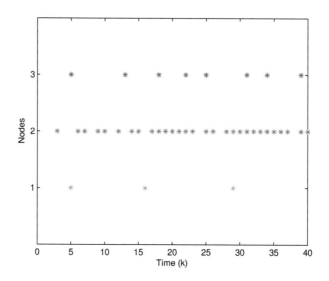

FIGURE 9.2: The triggering sequences for **OP2** (*Case 1*).

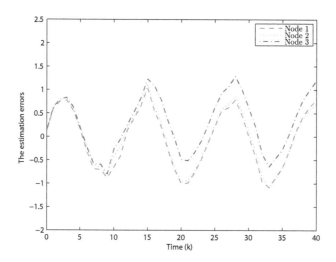

FIGURE 9.3: The estimation errors of $x_k^{(1)}$ for **OP1** (*Case 1*).

state entries $x_k^{(1)}$, $x_k^{(2)}$ are shown in Figs. 9.3–9.6. Generally, the adoption of the scheme developed in Corollary 9.1 results in a better filtering accuracy as expected, since it can be seen that the values of estimation errors shown in

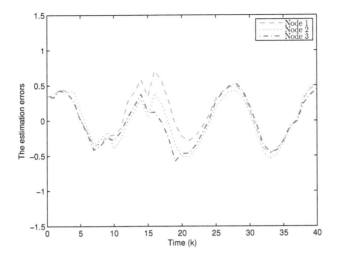

FIGURE 9.4: The estimation errors of $x_k^{(2)}$ for **OP1** (*Case 1*).

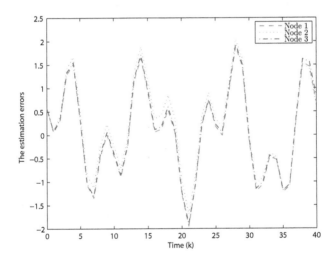

FIGURE 9.5: The estimation errors of $x_k^{(1)}$ for **OP2** (*Case 1*).

Fig. 9.3 and Fig. 9.4 are smaller than those shown in Fig. 9.5 and Fig. 9.6, respectively.

In order to figure out the impact from the saturation bound on filtering performance, we now consider the proposed *Case 2* expressed by (9.53). The

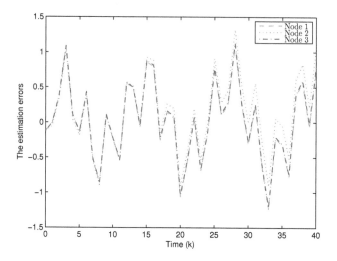

FIGURE 9.6: The estimation errors of $x_k^{(2)}$ for **OP2** (*Case 1*).

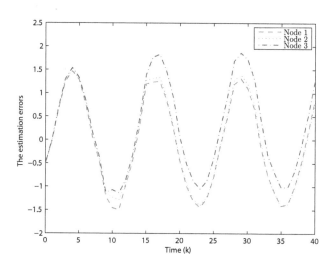

FIGURE 9.7: The estimation errors of $x_k^{(1)}$ for **OP1** (*Case 2*).

simulation results are shown in Fig. 9.7 and Fig. 9.8, where the figures of the estimation errors are presented. We can see that filtering performance can be improved if the saturation bound becomes larger.

In summary, the proposed design technique for the desired distributed event-based filter offers much flexibility in making trade-offs between the two

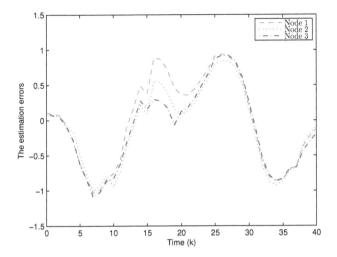

FIGURE 9.8: The estimation errors of $x_k^{(2)}$ for **OP1** (*Case 2*).

essential requirements (i.e., filtering performance and triggering frequency) and therefore provides engineers with an effective methodology of achieving the balance between accuracy and cost in practical applications.

9.4 Summary

In this chapter, the distributed event-based set-membership filtering problem has been addressed for a class of discrete nonlinear time-varying systems subject to unknown but bounded noises and sensor saturations over sensor networks. A novel event-triggering communication mechanism has been proposed for the sake of reducing the sensor data transmission rate and energy consumption. By means of the recursive linear matrix inequalities approach, the sufficient conditions have been established for the existence of the desired distributed event-triggering filter. With the established framework, two optimization problems have been discussed to demonstrate the flexibility of the proposed methodology in making trade-offs between accuracy and cost. Finally, a numerical simulation example has been exploited to verify the effectiveness of the distributed event-triggering filtering strategy.

10

Variance-Constrained Distributed Filtering for Time-varying Systems with Multiplicative Noises and Deception Attacks over Sensor Networks

The rapid development of microelectronic technologies over the past few decades has boosted the utilization of networks which consist of a large number of devices implemented distributively for sensing and communication as well as actuating. A quintessential example is the sensor network which has found wide applications ranging from various industrial branches to critical infrastructures such as military facilities and power grids. In particular, the state estimation or filtering problems over sensor networks have posed many emerging challenges, which have attracted ever-increasing research attention within the signal processing and control community.

So far, considerable effort has been devoted to the investigation of the distributed filtering problems and many strategies have been developed based on the Kalman filtering theory or the $\mathcal{H}_\infty$ filtering theory. However, as is well known, the Kalman filtering technique requires an assumption of Gaussian distributions for the process and measurement noises, while the $\mathcal{H}_\infty$ theory can be utilized on the occasion when the disturbances are assumed to have bounded energy. However, in many real-world engineering practices, due to a variety of reasons (e.g., man-made electromagnetic interference), it is much more appropriate to model the disturbances/noises as signals that are unknown but bounded in certain sets rather than as Gaussian noises or energy-bounded disturbances. Obviously, in such a case, the aforementioned conventional techniques based on Kalman filtering or $\mathcal{H}_\infty$ filtering frameworks are no longer effective. Consequently, the filtering problems for systems subject to the so-called unknown but bounded noises have excited tremendous fascination among researchers as well as engineers within the signal processing community. Nevertheless, in the general context of *sensor networks*, little progress has been made on the corresponding filtering problems owing probably to the difficulty in quantifying the filtering performance with respect to the unknown but bounded noises as well as the complexity which stems from the coupling between communication topology and system dynamics.

Along with the pervasive utilization of open yet unprotected communication networks, the sensor networks are vulnerable to cyber threat. As a result, the security of networks, which is of utmost importance in network-related systems, has provoked increasing research interest and a multitude of results have been reported in the literature. In general, there are mainly two types of cyber attacks which can affect the system's behavior directly or through feedback, namely, denial-of-Service (DoS) attacks and deception attacks. The DoS attack deteriorates system performance by preventing information from reaching the destination, while the deception attack aims at manipulating the system toward the adversaries' desired behaviors by tampering with control actions or system measurements. To tackle the filtering/control problems for systems under cyber attacks, several approaches have been developed, including linear programming, the linear matrix inequality method, and the game theory approach, to name but a few key ones. However, when it comes to distributed filtering issues over sensor networks, the corresponding results have been scattered. The difficulty probably lies in the lack of appropriate attack models which, on one hand, could comprehensively reflect engineering practice, and on the other, can be handled systematically within existing frameworks.

In response to the above discussion, it is our objective in this chapter to design a distributed filter for the discrete time-varying systems with multiplicative noises, unknown but bounded disturbances and deception attacks such that the estimation error variance of each sensing node is constrained by a pre-specified upper-bound at each time instant. Note that the specific time-varying nature of the addressed system imposes substantial challenges in both performance analysis and filter design, not to mention the difficulties stemming from the coupling between the communication topology and the deception attacks, especially when the error variances are required to satisfy certain upper-bounds at each time step. Therefore, we shall make the first of a few attempts to develop new paradigms to solve the so-called variance-constrained distributed filtering problem subject to deception attacks over sensor networks.

The rest of this chapter is organized as follows: Section 10.1 formulates the variance-constrained distributed filter design problem for the discrete time-varying system subject to multiplicative noises, unknown but bounded disturbances and deception attacks. The main results are presented in Section 10.2 where a sufficient condition for the existence of the desired filter is given in terms of recursive linear matrix inequalities. Section 10.3 gives a numerical example. Section 10.4 is our conclusion.

10.1   Problem Formulation

In this chapter, it is assumed that the sensor network has N sensor nodes which are distributed in the space according to a specific interconnection topology characterized by a directed graph $\mathscr{G} = (\mathscr{V}, \mathscr{M}, \mathscr{L})$, where $\mathscr{V} = \{1, 2, ..., N\}$ denotes the set of sensor nodes, $\mathscr{M} \subseteq \mathscr{V} \times \mathscr{V}$ is the set of edges, and $\mathscr{L} = [\theta_{ij}]_{N \times N}$ is the nonnegative adjacency matrix associated with the edges of the graph; i.e., $\theta_{ij} > 0 \Leftrightarrow$ edge $(i, j) \in \mathscr{M}$, which means that there is information transmission from sensor j to sensor i. If $(i, j) \in \mathscr{M}$, then node j is called one of the neighbors of node i. Also, we assume that $\theta_{ii} = 1$ for all $i \in \mathscr{V}$, and therefore, (i, i) can be regarded as an additional edge. The set of neighbors of node $i \in \mathscr{V}$ plus the node itself is denoted by $\mathscr{N}_i \triangleq \{j \in \mathscr{V} | (i, j) \in \mathscr{M}\}$.

System model

Consider a discrete time-varying system described by

$$x_{k+1} = \left(A_k + \sum_{l=1}^{q} \alpha_{l,k} A_{l,k}\right) x_k + B_k u_k + D_k w_k, \qquad (10.1)$$

with the measurements from N sensors given by

$$y_{i,k} = C_{i,k} x_k + E_{i,k} v_k, \quad i = 1, 2, \ldots, N \qquad (10.2)$$

where $x_k \in \mathbb{R}^n$, $u_k \in \mathbb{R}^m$ and $y_{i,k} \in \mathbb{R}^p$ are, respectively, the state, the known input and the measurement output of sensor i; $\alpha_{l,k} \in \mathbb{R}$ $(l = 1, 2, \ldots, q)$ are sequences of uncorrelated zero-mean Gaussian noises with unitary covariances; $w_k \in \mathbb{R}^\omega$ and $v_k \in \mathbb{R}^\nu$ represent the unknown but bounded process and measurement disturbances, respectively; A_k, B_k, D_k, $C_{i,k}$ and $E_{i,k}$ are real-valued time-varying matrices of compatible dimensions. The following definitions as well as assumptions are needed for further development.

Definition 10.1 *Let Ψ_1 and Ψ_2 be some real matrices with $\Psi \triangleq \Psi_2 - \Psi_1 > 0$. A nonlinearity $\varphi(\cdot)\colon \mathbb{R}^n \mapsto \mathbb{R}^n$ is said to satisfy the sector condition with respect to Ψ_1 and Ψ_2 if*

$$\left(\varphi(\varepsilon) - \Psi_1\varepsilon\right)^{\mathrm{T}}\left(\varphi(\varepsilon) - \Psi_2\varepsilon\right) \leq 0, \quad \forall \varepsilon \in \mathbb{R}^n. \qquad (10.3)$$

In this case, the sector-bounded nonlinearity $\varphi(\cdot)$ is said to belong to the sector $[\Psi_1, \Psi_2]$.

Definition 10.2 *A bounded ellipsoid $\mathscr{E}(c, P, n)$ of $\mathbb{R}^n$ with a nonempty interior can be defined by*

$$\mathscr{E}(c, P, n) \triangleq \{x \in \mathbb{R}^n : (x - c)^{\mathrm{T}} P^{-1}(x - c) \leq 1\} \qquad (10.4)$$

where $c \in \mathbb{R}^n$ is the center of $\mathscr{E}(c, P, n)$ and $P > 0$ is a positive definite matrix that specifies the ellipsoid's shape and orientation.

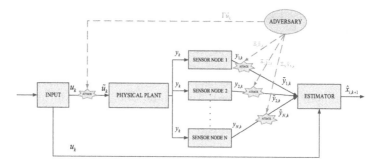

FIGURE 10.1: Deception attack model.

Assumption 10.1 *The unknown but bounded noises w_k and v_k are confined to the following specified ellipsoids:*

$$w_k \in \mathscr{E}(0, W_k, \omega) \triangleq \{w_k \in \mathbb{R}^\omega : w_k^{\mathrm{T}} W_k^{-1} w_k \leq 1\}$$
$$v_k \in \mathscr{E}(0, V_k, \nu) \triangleq \{v_k \in \mathbb{R}^\nu : v_k^{\mathrm{T}} V_k^{-1} v_k \leq 1\}$$

(10.5)

where $W_k > 0$ and $V_k > 0$ are known positive definite matrices of appropriate dimensions.

Remark 10.1 *In many real-world applications of signal processing as well as stochastic control, due to various reasons such as man-made electromagnetic interferences [60], it is more appropriate to model the disturbance noises as unknown bounded disturbances than stochastic noises (usually Gaussian) or energy-bounded noises. In (10.5), the disturbances w_k and v_k have been assumed to be unknown but residing within certain specified ellipsoidal sets. It should be pointed out that, in this instance, traditional techniques (e.g. Kalman filtering method or $\mathcal{H}_\infty$ design approach) cannot be directly utilized to examine the corresponding filtering problems against the unknown but bounded disturbances.*

Deception attack model

In this chapter, we investigate the following deception attack scenario. Attempting to deteriorate the filtering performance, the adversary injects certain deception signals into the true signals of the control input u_k and the measurement outputs $y_{i,k}$ during the process of data transmission through the communication networks. Such an attack scenario can be illustrated by Fig. 10.1.

Before giving the deception attack model, we make some further assumptions on the system knowledge that is possessed by the adversary for implementing a successful attack. In this chapter, it is assumed that the adversary has sufficient resources and adequate knowledge to arrange a successful attack [156]. Specifically, the adversary, in the first place, knows the accurate values of control input u_k and the measurement output $y_{i,k}$ in real time, and in the second place, has the ability to modify the true values of u_k and $y_{i,k}$ to arbitrary ones. Moreover, the attacks are arranged in a coordinated fashion where the deception signals are injected into each communication channel simultaneously to maximize the impact to the plant/estimator [200].

The signals used by the adversary for the deception attacks are generated as follows:

$$\begin{cases} \vec{u}_k = -u_k + \delta_k \\ \vec{y}_{i,k} = -y_{i,k} + \vartheta_{i,k}, \quad i = 1, 2, \ldots, N \end{cases} \tag{10.6}$$

where δ_k and $\vartheta_{i,k}$ $(i = 1, 2, \ldots, N)$ are the unknown but bounded signals belonging to the following ellipsoids:

$$\begin{aligned} \delta_k \in \mathscr{E}(0, S_k, m) &\triangleq \{\delta_k \in \mathbb{R}^m : \delta_k^{\mathrm{T}} S_k^{-1} \delta_k \leq 1\}, \\ \vartheta_{i,k} \in \mathscr{E}(0, R_{i,k}, p) &\triangleq \{\vartheta_{i,k} \in \mathbb{R}^p : \vartheta_{i,k}^{\mathrm{T}} R_{i,k}^{-1} \vartheta_{i,k} \leq 1\} \end{aligned} \tag{10.7}$$

with S_k and $R_{i,k}$ $(i = 1, 2, \ldots, N)$ being positive definite matrices of compatible dimensions.

Remark 10.2 *In* (10.7), δ_k *and* $\vartheta_{i,k}$ $(i = 1, 2, \ldots, N)$, *which have been assumed to be unknown but confined to certain ellipsoidal sets, are used by the adversary to generate the deception attack signals. It should be noted that* δ_k *and* $\vartheta_{i,k}$ *have similar forms with the process noise* w_k *and the measurement noise* v_k, *and are therefore difficult to be distinguished by the detectors. On the other hand, the most widely implemented attack detector in the practical applications, namely, the* χ^2 *detector, is only effective when the noises obey Gaussian distribution [15]. As such, the utilization of unknown but bounded signals could help to pass through the* χ^2 *detector. In other words, from the adversary's perspective, it is practically reasonable to constrain the malicious signals* δ_k *and* $\vartheta_{i,k}$ $(i = 1, 2, \ldots, N)$ *within given ellipsoidal sets.*

Remark 10.3 *In engineering practice, attack detectors are categorized as a software barrier, and there are some other "hard" physical constraints that the adversary would need to face. Such physical constraints include device saturations, bandwidth limitations, channel fading and signal quantizations [50]. The kinds of hardware constraints should be taken into consideration if we are to establish a comprehensive yet realistic deception attack model. On the other hand, such constraints inevitably bring in new challenges that demand new techniques in analyzing the performance and designing the filters.*

Based on the discussions in Remark 10.3 and from Fig. 10.1, we can reformulate the actual control input $\tilde{u}_k$ (sent to the plant) and the actual measurement outputs $\tilde{y}_{i,k}$ (fed to the estimator) by

$$\begin{cases} \tilde{u}_k = u_k + \Gamma \vec{u}_k \\ \tilde{y}_{i,k} = y_{i,k} + \Xi_i \vec{y}_{i,k}, \quad i = 1, 2, \ldots, N \end{cases} \tag{10.8}$$

where the matrices Γ and Ξ_i represent the physical constraints imposed on the attack signals and are assumed to be of the following forms:

$$\begin{cases} \Gamma = \mathrm{diag}\{\gamma_1, \gamma_2, \ldots, \gamma_m\}, \\ \Xi_i = \mathrm{diag}\{\xi_{i,1}, \xi_{i,2}, \ldots, \xi_{i,p}\}, \quad i = 1, 2, \ldots, N. \end{cases} \tag{10.9}$$

Here, the entries of Γ and Ξ_i have upper- and lower-bounds that are expressed as follows:

$$\begin{cases} 0 \leq \underline{\gamma}_j \leq \gamma_j \leq \bar{\gamma}_j < \infty, \quad j = 1, 2, \ldots, m \\ 0 \leq \underline{\xi}_{i,s} \leq \xi_{i,s} \leq \bar{\xi}_{i,s} < \infty, \quad s = 1, 2, \ldots, p \end{cases} \tag{10.10}$$

where $0 \leq \underline{\gamma}_j < 1$ and $\bar{\gamma}_j \geq 1$ are known scalars representing the lower- and upper-bounds on γ_j, and $0 \leq \underline{\xi}_{i,s} < 1$ and $\bar{\xi}_{i,s} \geq 1$ are known scalars describing the lower- and upper-bounds on $\xi_{i,s}$, respectively.

Remark 10.4 *We now take the matrix Γ as an example to illustrate how the upper- and lower-bounds on γ_j affect the behavior of the deception attack signal $\vec{u}_k$ (the impact on $\vec{y}_{i,k}$ from $\xi_{i,s}$ can be analyzed similarly). Specifically, when $\gamma_j = 1$, it means that the jth entry of the deception signal $\vec{u}_k$ can be injected correctly into the corresponding true u_k as the adversary plans; otherwise $\vec{u}_k$ might be unexpectedly (from the attacker's perspective) degraded $(0 \leq \gamma_j < 1)$ or amplified $(\gamma_j > 1)$. In this sense, the model (10.8)–(10.10) offers a comprehensive and realistic means of reflecting the influence on the attacks resulting from the physical constraints as well as the network-induced complexities.*

By denoting

$$\underline{\Gamma} \triangleq \mathrm{diag}\{\underline{\gamma}_1, \underline{\gamma}_2, \ldots, \underline{\gamma}_m\}, \quad \bar{\Gamma} \triangleq \mathrm{diag}\{\bar{\gamma}_1, \bar{\gamma}_2, \ldots, \bar{\gamma}_m\},$$

$$\underline{\Xi}_i \triangleq \mathrm{diag}\{\underline{\xi}_{i,1}, \underline{\xi}_{i,2}, \ldots, \underline{\xi}_{i,p}\}, \quad \bar{\Xi}_i \triangleq \mathrm{diag}\{\bar{\xi}_{i,1}, \bar{\xi}_{i,2}, \ldots, \bar{\xi}_{i,p}\},$$

we can rewrite (10.10) into the following compact forms:

$$\begin{cases} \underline{\Gamma} \leq \Gamma \leq \bar{\Gamma}, \\ \underline{\Xi}_i \leq \Xi_i \leq \bar{\Xi}_i, \quad i = 1, 2, \ldots, N. \end{cases} \tag{10.11}$$

In what follows, for the convenience of later derivation, we divide the deception attack signals $\Gamma \vec{u}_k$ and $\Xi_i \vec{y}_{i,k}$ as follows:

$$\begin{cases} \Gamma \vec{u}_k = \underline{\Gamma} \vec{u}_k + \varphi(\vec{u}_k), \\ \Xi_i \vec{y}_{i,k} = \underline{\Xi}_i \vec{y}_{i,k} + \psi_i(\vec{y}_{i,k}), \quad i = 1, 2, \ldots, N. \end{cases} \tag{10.12}$$

It then can be easily checked that

$$\begin{cases} \varphi^{\mathrm{T}}(\vec{u}_k)\big(\varphi(\vec{u}_k) - \tilde{\Gamma}\vec{u}_k\big) \leq 0, \\ \psi_i^{\mathrm{T}}(\vec{y}_{i,k})\big(\psi_i(\vec{y}_{i,k}) - \tilde{\Xi}_i\vec{y}_{i,k}\big) \leq 0, \quad i = 1, 2, \ldots, N \end{cases} \tag{10.13}$$

where $\tilde{\Gamma} \triangleq \bar{\Gamma} - \underline{\Gamma} > 0$ and $\tilde{\Xi}_i \triangleq \bar{\Xi}_i - \underline{\Xi}_i > 0$ are positive definite matrices. Clearly, it follows from Definition 10.1 that $\varphi(\vec{u}_k)$ and $\psi_i(\vec{y}_{i,k})$ are vector-valued nonlinear functions satisfying the sector condition and belonging to the sectors $[0, \tilde{\Gamma}]$ and $[0, \tilde{\Xi}_i]$ $(i = 1, 2, \ldots, N)$, respectively.

Design objective

On account of the deception attacks discussed above, the original system (10.1)–(10.2) should be reformulated by

$$
\begin{cases}
x_{k+1} = \left(A_k + \sum_{l=1}^{q} \alpha_{l,k} A_{l,k} \right) x_k + B_k \tilde{u}_k + D_k w_k, \\
\tilde{u}_k = u_k + \Gamma \vec{u}_k, \\
\tilde{y}_{i,k} = y_{i,k} + \Xi_i \vec{y}_{i,k}, \quad i = 1, 2, \ldots, N.
\end{cases}
\tag{10.14}
$$

For the system (10.14), at each sensing node i $(i = 1, 2, \ldots, N)$, the following filter structure is adopted:

$$
\hat{x}_{i,k+1} = G_{i,k} \hat{x}_{i,k} + B_k u_k + \sum_{j \in \mathcal{N}_i} \theta_{ij} K_{ij,k} (\tilde{y}_{i,k} - C_{i,k} \hat{x}_{i,k})
\tag{10.15}
$$

where $\hat{x}_{i,k} \in \mathbb{R}^n$ is the estimate of the state x_k based on the ith sensing node, and the matrices $G_{i,k}$ and $K_{ij,k}$ $(i, j = 1, 2, \ldots, N)$ are the filter parameters to be determined.

Assumption 10.2 *The initial state x_0 of the system (10.14) and its estimate values from each sensing node, namely, $\hat{x}_{i,0}$ $(i = 1, 2, \ldots, N)$, satisfy:*

$$
\mathbb{E} \left\{ (x_0 - \hat{x}_{i,0})(x_0 - \hat{x}_{i,0})^{\mathrm{T}} \right\} \leq \Phi_0
\tag{10.16}
$$

where $\Phi_0 > 0$ is a known positive definite matrix.

The distributed filtering problem under investigation is to estimate the state of the system (10.1) using a network of filters connected according to the graph $\mathscr{G}$ with the guarantee of variance constraints on the estimation errors. Specifically, the objective of this chapter is twofold. For the system (10.1)–(10.2) subject to the deception attacks (10.6), let the communication graph $\mathscr{G}$ and the sequence of positive definite matrices $\{\Phi_k\}_{k \geq 0}$ (prespecified constraints on the estimation error variance) be given. It is our first aim to design the sequences of filtering parameters $\{G_{i,k}\}_{k \geq 0}$ and $\{K_{ij,k}\}_{k \geq 0}$ in (10.15) subject to the given couple $(\mathscr{G}, \{\Phi_k\}_{k \geq 0})$ such that the following inequalities are satisfied for all $k \geq 0$:

$$
\mathbb{E} \left\{ (x_k - \hat{x}_{i,k})(x_k - \hat{x}_{i,k})^{\mathrm{T}} \right\} \leq \Phi_k, \quad i = 1, 2, \ldots, N.
\tag{10.17}
$$

Second, within the proposed framework, an optimization problem will be considered for minimizing Φ_k in the sense of matrix trace at each time instant to ensure the locally optimal filtering performance. This problem will be referred to as a variance-constrained distributed filtering problem subject to deception attacks.

10.2   Distributed Filter Design

In this section, we are going to design a distributed filter of form (10.15) for system (10.1)–(10.2) subject to multiplicative noises, unknown but bounded disturbances and deception attacks. A sufficient condition for the existence of the desired filter will be formulated in terms of a set of recursive linear matrix inequalities (RLMIs). First, two lemmas which are useful for our subsequent development are introduced as follows.

Lemma 10.1 *(Schur Complement Lemma) Given constant matrices $\mathcal{S}_1, \mathcal{S}_2, \mathcal{S}_3$ where $\mathcal{S}_1 = \mathcal{S}_1^{\mathrm{T}}$ and $0 < \mathcal{S}_2 = \mathcal{S}_2^{\mathrm{T}}$, then $\mathcal{S}_1 + \mathcal{S}_3^{\mathrm{T}} \mathcal{S}_2^{-1} \mathcal{S}_3 < 0$ if and only if*

$$\begin{bmatrix} \mathcal{S}_1 & \mathcal{S}_3^{\mathrm{T}} \\ \mathcal{S}_3 & -\mathcal{S}_2 \end{bmatrix} < 0, \quad \text{or} \quad \begin{bmatrix} -\mathcal{S}_2 & \mathcal{S}_3 \\ \mathcal{S}_3^{\mathrm{T}} & \mathcal{S}_1 \end{bmatrix} < 0. \qquad (10.18)$$

Lemma 10.2 *(S-procedure [18]) Let $\kappa_0(\cdot), \kappa_1(\cdot), \dots, \kappa_s(\cdot)$ be quadratic functions of the variable $\zeta \in \mathbb{R}^n$: $\kappa_j(\zeta) \triangleq \zeta^{\mathrm{T}} T_j \zeta \ (j = 0, 1, \dots, s)$, where $T_j^{\mathrm{T}} = T_j$. If there exist $\tau_1 \geq 0, \ \dots, \ \tau_s \geq 0$ such that $T_0 - \sum_{j=1}^{s} \tau_j T_j \leq 0$, then the following is true:*

$$\kappa_1(\zeta) \leq 0, \dots, \kappa_s(\zeta) \leq 0 \Longrightarrow \kappa_0(\zeta) \leq 0. \qquad (10.19)$$

From the system (10.14) and the filter (10.15), for sensing node i $(i = 1, 2, \dots, N)$, the one-step ahead estimation error is calculated by

$$x_{k+1} - \hat{x}_{i,k+1}$$
$$= \left(A_k + \sum_{l=1}^{q} \alpha_{l,k} A_{l,k} \right) x_k - B_k \underline{\Gamma} u_k + B_k \underline{\Gamma} \delta_k + B_k \varphi(\vec{u}_k)$$
$$+ D_k w_k - G_{i,k} \hat{x}_{i,k} - \sum_{j \in \mathcal{N}_i} \theta_{ij} K_{ij,k}(I - \underline{\Xi}_i) C_{i,k} x_k$$
$$- \sum_{j \in \mathcal{N}_i} \theta_{ij} K_{ij,k}(I - \underline{\Xi}_i) E_{i,k} v_k - \sum_{j \in \mathcal{N}_i} \theta_{ij} K_{ij,k} \underline{\Xi}_i \vartheta_{i,k}$$
$$- \sum_{j \in \mathcal{N}_i} \theta_{ij} K_{ij,k} \psi_i(\vec{y}_{i,k}) + \sum_{j \in \mathcal{N}_i} \theta_{ij} K_{ij,k} C_{i,k} \hat{x}_{i,k}. \qquad (10.20)$$

For simplicity of further development, we denote

$$\mathcal{T}_{\lambda,i} \triangleq [\underbrace{0 \cdots 0}_{i-1} \ I_\lambda \ \underbrace{0 \cdots 0}_{N-i}],$$
$$\Omega_{\lambda,i} \triangleq \mathcal{T}_{\lambda,i}^{\mathrm{T}} \mathcal{T}_{\lambda,i}(I_N \otimes I_\lambda), \quad \lambda = \{n, q, p\}, \ i = 1, \dots, N.$$

Denoting $\tilde{x}_{i,k} \triangleq x_k - \hat{x}_{i,k}$, we can rewrite (10.20) into a compact form as follows:

$$
\begin{aligned}
\tilde{x}_{k+1} =& (\mathcal{A}_k + \bar{\alpha}_k \bar{\mathcal{A}}_k)\zeta_k - \mathcal{B}_k \mu_k + \mathcal{B}_k \tilde{\delta}_k + \bar{\mathcal{B}}_k \tilde{\varphi}_k \\
& + \mathcal{D}_k \tilde{w}_k - \mathcal{G}_k \hat{x}_k - \mathcal{K}_k (I - \Xi)\mathcal{C}_k \zeta_k \\
& - \mathcal{K}_k (I - \Xi)\mathcal{E}_k \tilde{v}_k - \mathcal{K}_k \Xi \vartheta_k \\
& - \mathcal{K}_k \psi_k + \mathcal{K}_k \mathcal{C}_k \hat{x}_k
\end{aligned}
\tag{10.21}
$$

where

$$
\begin{aligned}
& \tilde{x}_k \triangleq \mathrm{col}_N\{\tilde{x}_{i,k}\}, \ \ \zeta_k \triangleq \mathrm{col}_N\{x_k\}, \ \ \hat{x}_k \triangleq \mathrm{col}_N\{\hat{x}_{i,k}\}, \\
& \mu_k \triangleq \mathrm{col}_N\{u_k\}, \ \ \tilde{\delta}_k \triangleq \mathrm{col}_N\{\delta_k\}, \ \ \tilde{\varphi}_k \triangleq \mathrm{col}_N\{\varphi(\vec{u}_k)\}, \\
& \tilde{w}_k \triangleq \mathrm{col}_N\{w_k\}, \ \ \tilde{v}_k \triangleq \mathrm{col}_N\{v_k\}, \ \ \vartheta_k \triangleq \mathrm{col}_N\{\vartheta_{i,k}\}, \\
& \psi_k \triangleq \mathrm{col}_N\{\psi_i(\vec{y}_{i,k})\}, \ \ \mathcal{A}_k \triangleq \mathrm{diag}_N\{A_k\}, \\
& \tilde{A}_k \triangleq \mathrm{diag}_N\left\{\sum_{l=1}^{q} A_{l,k}\right\}, \ \ \alpha_k \triangleq \begin{bmatrix} \alpha_{1,k}I & \cdots & \alpha_{q,k}I \end{bmatrix}, \\
& \bar{A}_k \triangleq \begin{bmatrix} A_{1,k}^\mathrm{T} & A_{2,k}^\mathrm{T} & \cdots & A_{q,k}^\mathrm{T} \end{bmatrix}^\mathrm{T}, \ \ \bar{\alpha}_k \triangleq \mathrm{diag}_N\{\alpha_k\}, \\
& \bar{\mathcal{A}}_k \triangleq \mathrm{diag}_N\{\bar{A}_k\}, \ \ \mathcal{B}_k \triangleq \mathrm{diag}_N\{B_k \Gamma\}, \ \ \bar{\mathcal{B}}_k \triangleq \mathrm{diag}_N\{B_k\}, \\
& \mathcal{C}_k \triangleq \mathrm{diag}_N\{C_{i,k}\}, \ \ \mathcal{D}_k \triangleq \mathrm{diag}_N\{D_{i,k}\}, \ \ \mathcal{E}_k \triangleq \mathrm{diag}_N\{E_{i,k}\}, \\
& \mathcal{G}_k \triangleq \mathrm{diag}_N\{G_{i,k}\}, \ \ \Xi \triangleq \mathrm{diag}_N\{\Xi_i\}, \ \ \mathcal{K}_k \triangleq [\theta_{ij}K_{ij,k}]_{N \times N}.
\end{aligned}
$$

Noticing that when $j \notin \mathcal{N}_i$, $\theta_{ij} = 0$, we know that $\mathcal{K}_k$ is a sparse matrix which can be described by

$$
\mathcal{K}_k \in \mathscr{Q}_{n \times m}
\tag{10.22}
$$

where $\mathscr{Q}_{n \times m} \triangleq \{\mathcal{Q} = [Q_{ij}] \in \mathbb{R}^{nN \times mN} | Q_{ij} \in \mathbb{R}^{n \times m}, Q_{ij} = 0 \text{ if } j \notin \mathcal{N}_i\}$.

The following theorem presents a sufficient condition for the existence of the desired distributed filter by RLMI approach.

Theorem 10.1 *For the system* (10.1)–(10.2) *subject to the deception attacks* (10.6), *let the network topology $\mathscr{G}$ and the prespecified sequence of variance constraints $\{\Phi_k\}_{k \geq 0}$ be given. The design objective* (10.17) *is achieved if there exist sequences of real-valued matrices $\{\mathcal{G}_k\}_{k \geq 0}$ and $\{\mathcal{K}_k\}_{k \geq 0}$ ($\mathcal{K}_k \in \mathscr{Q}_{n \times m}$), sequences of positive definite matrices $\{\tilde{S}_k\}_{k \geq 0}$ and $\{\tilde{R}_{i,k}\}_{k \geq 0}$, sequences of non-negative scalars $\{\tau_{1,k}\}_{k \geq 0}$, $\{\tau_{2,k}\}_{k \geq 0}$, $\{\tau_{3,k}\}_{k \geq 0}$, $\{\tau_{4,k}\}_{k \geq 0}$, $\{\epsilon_{i,k}\}_{k \geq 0}$, $\{\varrho_{i,k}\}_{k \geq 0}$ and $\{\rho_{i,k}\}_{k \geq 0}$ ($i = 1, 2, \ldots, N$) satisfying the following N RLMIs:*

$$
\begin{bmatrix} -\Delta_k & * & * \\ \mathcal{T}_{n,i}\bar{\Lambda}_k & -\Phi_{k+1} & * \\ \mathcal{T}_{n,i}\hat{\Lambda}_k & 0 & -\Phi_{k+1} \end{bmatrix} \leq 0, \qquad i = 1, 2, \ldots, N
\tag{10.23}
$$

where

$$\Delta_k \triangleq \tau_{4,k}\Pi_k + \sum_{i=1}^{N} \rho_{i,k}\Upsilon_{i,k}$$

$$+ \operatorname{diag}\Big\{ 1 - \sum_{i=1}^{N} (\epsilon_{i,k} + \varrho_{i,k}) - \tau_{1,k} - \tau_{2,k} - \tau_{3,k},$$

$$\sum_{i=1}^{N} \epsilon_{i,k}\mathcal{T}_{q,i}^{\mathrm{T}}\mathcal{T}_{q,i}, \tilde{S}_k, \sum_{i=1}^{N}\mathcal{T}_{p,i}^{\mathrm{T}}\tilde{R}_{i,k}\mathcal{T}_{p,i},$$

$$\tau_{2,k}W_k^{-1}, \tau_{3,k}V_k^{-1}, 0, 0 \Big\}, \tag{10.24}$$

$$\bar{\Pi}_k \triangleq \begin{bmatrix} u_k^{\mathrm{T}}\tilde{\Gamma} \\ 0 \\ -\tilde{\Gamma} \\ 0 \\ 0 \\ 0 \\ 2I_m \\ 0 \end{bmatrix}, \quad \bar{\Upsilon}_{i,k} \triangleq \begin{bmatrix} \hat{x}_k^{\mathrm{T}}\Omega_{n,i}\operatorname{diag}_N\{C_{i,k}^{\mathrm{T}}\tilde{\Xi}_i\}\Omega_{p,i} \\ \Omega_{q,i}\operatorname{diag}_N\{P_k^{\mathrm{T}}C_{i,k}^{\mathrm{T}}\tilde{\Xi}_i\}\Omega_{p,i} \\ 0 \\ -\Omega_{p,i}\operatorname{diag}_N\{\tilde{\Xi}_i\}\Omega_{p,i} \\ 0 \\ \mathcal{T}_{p,i}\operatorname{diag}_N\{E_{i,k}^{\mathrm{T}}\tilde{\Xi}_i\}\Omega_{p,i} \\ 0 \\ 2\Omega_{p,i}I_{pN}\Omega_{p,i} \end{bmatrix},$$

$$\Pi_k \triangleq \frac{1}{2}\begin{bmatrix} 0 & 0 & 0 & 0 & 0 & 0 & \bar{\Pi}_k & 0 \end{bmatrix}, \tag{10.25}$$

$$\Upsilon_{i,k} \triangleq \frac{1}{2}\begin{bmatrix} 0 & 0 & 0 & 0 & 0 & 0 & 0 & \bar{\Upsilon}_{i,k} \end{bmatrix}, \tag{10.26}$$

$$\mathscr{I}_m \triangleq \mathbf{1}_N \otimes I_m, \quad \mathscr{I}_\omega \triangleq \mathbf{1}_N \otimes I_\omega, \quad \mathscr{I}_\nu \triangleq \mathbf{1}_N \otimes I_\nu, \tag{10.27}$$

$$\bar{\Lambda}_{11} \triangleq \Big(\mathcal{A}_k - \mathcal{G}_k + \mathcal{K}_k\Xi\mathcal{C}_k\Big)\hat{x}_k - \mathcal{B}_k\mu_k,$$

$$\bar{\Lambda}_{12} \triangleq \mathcal{A}_k\mathcal{P}_k - \mathcal{K}_k(I - \Xi)\mathcal{C}_k\mathcal{P}_k,$$

$$\bar{\Lambda}_k \triangleq \begin{bmatrix} \bar{\Lambda}_{11} & \bar{\Lambda}_{12} & \mathcal{B}_k\mathscr{I}_m & -\mathcal{K}_k\Xi & \mathcal{D}_k\mathscr{I}_\omega \end{bmatrix}$$
$$\begin{bmatrix} -\mathcal{K}_k(I - \Xi)\mathcal{E}_k\mathscr{I}_\nu & \bar{\mathcal{B}}_k\mathscr{I}_m & -\mathcal{K}_k \end{bmatrix}, \tag{10.28}$$

$$\hat{\Lambda}_k \triangleq \begin{bmatrix} \tilde{\mathcal{A}}_k\hat{x}_k & \tilde{\mathcal{A}}_k\mathcal{P}_k & 0 & 0 & 0 & 0 & 0 & 0 \end{bmatrix} \tag{10.29}$$

with P_k being a factorization of Φ_k (i.e., $\Phi_k = P_kP_k^{\mathrm{T}}$) and $\mathcal{P}_k \triangleq \operatorname{diag}_N\{P_k\}$. Moreover, the parameters S_k and $R_{i,k}$ can be computed by $S_k = \tau_{1,k}\tilde{S}_k^{-1}$ and $R_{i,k} = \varrho_{i,k}\tilde{R}_{i,k}^{-1}$.

Proof *We prove Theorem 10.1 by induction which can be divided into two steps, namely, the initial step and the inductive step.*

Initial step. For $k = 0$, it can be known directly from Assumption 10.2 that the initial value of the state x_0 and its estimates $\hat{x}_{i,0}$ ($i = 1, 2, \ldots, N$) satisfy

$$\mathbb{E}\big\{(x_0 - \hat{x}_{i,0})(x_0 - \hat{x}_{i,0})^{\mathrm{T}}\big\} \le \Phi_0. \tag{10.30}$$

Inductive step. Given that at the time instant $k > 0$, the following is true:

$$\mathbb{E}\left\{(x_k - \hat{x}_{i,k})(x_k - \hat{x}_{i,k})^{\mathrm{T}}\right\} \le \Phi_k, \tag{10.31}$$

then we aim to, with the condition given in the theorem, demonstrate that the following set of inequalities holds for all i:

$$\mathbb{E}\left\{(x_{k+1} - \hat{x}_{i,k+1})(x_{k+1} - \hat{x}_{i,k+1})^{\mathrm{T}}\right\} \le \Phi_{k+1}. \tag{10.32}$$

Since the inequality (10.31) is true, it follows from [66] that there exists a sequence of vectors $z_{i,k} \in \mathbb{R}^q$ (with $\mathbb{E}\{z_{i,k}^{\mathrm{T}}z_{i,k}\} \le 1$) satisfying $x_k = \hat{x}_{i,k} + P_k z_{i,k}$ where $P_k \in \mathbb{R}^{n \times q}$ is a factorization of $\Phi_k = P_k P_k^{\mathrm{T}}$. Denoting $z_k \triangleq \mathrm{col}_N\{z_{i,k}\}$ and noticing $\mathcal{P}_k = \mathrm{diag}_N\{P_k\}$, we can further acquire that

$$\zeta_k = \hat{x}_k + \mathcal{P}_k z_k. \tag{10.33}$$

Now, substituting (10.33) into (10.21) yields

$$\begin{aligned}
\tilde{x}_{k+1} =& \left(\mathcal{A}_k + \bar{\alpha}_k \bar{\mathcal{A}}_k - \mathcal{G}_k - \mathcal{K}_k(I - \Xi)\mathcal{C}_k + \mathcal{K}_k\mathcal{C}_k\right)\hat{x}_k \\
& - \mathcal{B}_k\mu_k + \left((\mathcal{A}_k + \bar{\alpha}_k\bar{\mathcal{A}}_k)\mathcal{P}_k - \mathcal{K}_k(I - \Xi)\mathcal{C}_k\mathcal{P}_k\right)z_k \\
& + \mathcal{B}_k\mathscr{I}_m\delta_k + \bar{\mathcal{B}}_k\mathscr{I}_m\varphi_k + \mathcal{D}_k\mathscr{I}_\omega w_k \\
& - \mathcal{K}_k(I - \Xi)\mathcal{E}_k\mathscr{I}_\nu v_k - \mathcal{K}_k\Xi\vartheta_k - \mathcal{K}_k\psi_k.
\end{aligned} \tag{10.34}$$

We now define a vector as follows:

$$\beta_k \triangleq \begin{bmatrix} 1 \\ z_k \\ \delta_k \\ \vartheta_k \\ w_k \\ v_k \\ \varphi_k \\ \psi_k \end{bmatrix}. \tag{10.35}$$

Then, the one-step ahead estimation error $\tilde{x}_{k+1}$ in (10.34) is expressed by

$$\tilde{x}_{k+1} = \Lambda_k\beta_k \tag{10.36}$$

where

$$\begin{aligned}
\Lambda_{11} \triangleq& (\mathcal{A}_k + \bar{\alpha}_k\bar{\mathcal{A}}_k - \mathcal{G}_k + \mathcal{K}_k\Xi\mathcal{C}_k + \mathcal{G}_k)\hat{x}_k - \mathcal{B}_k\mu_k, \\
\Lambda_{12} \triangleq& (\mathcal{A}_k + \bar{\alpha}_k\bar{\mathcal{A}}_k)\mathcal{P}_k - \mathcal{K}_k(I - \Xi)\mathcal{C}_k\mathcal{P}_k + \mathcal{G}_k\mathcal{P}_k, \\
\Lambda_k \triangleq& \begin{bmatrix} \Lambda_{11} & \Lambda_{12} & \mathcal{B}_k\mathscr{I}_m & -\mathcal{K}_k\Xi & \mathcal{D}_k\mathscr{I}_\omega \\ & -\mathcal{K}_k(I - \Xi)\mathcal{E}_k\mathscr{I}_\nu & \bar{\mathcal{B}}_k\mathscr{I}_m & -\mathcal{K}_k \end{bmatrix}.
\end{aligned} \tag{10.37}$$

Next, we decompose the matrix Λ_k into a deterministic part $\bar{\Lambda}_k$ defined in

(10.28) and a stochastic part $\tilde{\Lambda}_k$ which contains the random variable $\bar{\alpha}_k$ as follows:

$$\Lambda_k = \bar{\Lambda}_k + \tilde{\Lambda}_k \tag{10.38}$$

where

$$\tilde{\Lambda}_k \triangleq \begin{bmatrix} \bar{\alpha}_k \bar{A}_k \hat{x}_k & \bar{\alpha}_k \bar{A}_k \mathcal{P}_k & 0 & 0 & 0 & 0 & 0 & 0 \end{bmatrix}.$$

Then, by taking into consideration the statistical properties of the random variable $\bar{\alpha}_k$, we have

$$\begin{aligned}
&\mathbb{E}\{\Lambda_k^{\mathrm{T}} \mathcal{T}_{n,i}^{\mathrm{T}} \Phi_{k+1}^{-1} \mathcal{T}_{n,i} \Lambda_k\} \\
=&\mathbb{E}\{(\bar{\Lambda}_k + \tilde{\Lambda}_k)^{\mathrm{T}} \mathcal{T}_{n,i}^{\mathrm{T}} \Phi_{k+1}^{-1} \mathcal{T}_{n,i} (\bar{\Lambda}_k + \tilde{\Lambda}_k)\} \\
=&\bar{\Lambda}_k^{\mathrm{T}} \mathcal{T}_{n,i}^{\mathrm{T}} \Phi_{k+1}^{-1} \mathcal{T}_{n,i} \bar{\Lambda}_k + \mathbb{E}\{\tilde{\Lambda}_k^{\mathrm{T}} \mathcal{T}_{n,i}^{\mathrm{T}} \Phi_{k+1}^{-1} \mathcal{T}_{n,i} \tilde{\Lambda}_k\} \\
=&\bar{\Lambda}_k^{\mathrm{T}} \mathcal{T}_{n,i}^{\mathrm{T}} \Phi_{k+1}^{-1} \mathcal{T}_{n,i} \bar{\Lambda}_k + \hat{\Lambda}_k^{\mathrm{T}} \mathcal{T}_{n,i}^{\mathrm{T}} \Phi_{k+1}^{-1} \mathcal{T}_{n,i} \hat{\Lambda}_k.
\end{aligned} \tag{10.39}$$

In the following, we shall proceed to deal with the constraints (10.13) imposed on the attack signals $\vec{u}_k$ and $\vec{y}_{i,k}$ ($i = 1, 2, \ldots, N$). It can be known from (10.13) that the nonlinear vector-valued functions $\varphi(\vec{u}_k)$ and $\psi_i(\vec{y}_{i,k})$ ($i = 1, 2, \ldots, N$) belong to the sectors $[0, \tilde{\Gamma}]$ and $[0, \tilde{\Xi}_i]$ ($i = 1, 2, \ldots, N$), respectively. As such, we can perform the following derivations:

$$\begin{aligned}
&\varphi^{\mathrm{T}}(\vec{u}_k)\big(\varphi(\vec{u}_k) - \tilde{\Gamma}\vec{u}_k\big) \le 0 \\
\Longleftrightarrow&\varphi^{\mathrm{T}}(\vec{u}_k)\varphi(\vec{u}_k) - \varphi^{\mathrm{T}}(\vec{u}_k)\tilde{\Gamma}\vec{u}_k \le 0 \\
\Longleftrightarrow&\varphi^{\mathrm{T}}(\vec{u}_k)\varphi(\vec{u}_k) - \varphi^{\mathrm{T}}(\vec{u}_k)\tilde{\Gamma}(-u_k + \delta_k) \le 0 \\
\Longleftrightarrow&\varphi^{\mathrm{T}}(\vec{u}_k)\varphi(\vec{u}_k) + \varphi^{\mathrm{T}}(\vec{u}_k)\tilde{\Gamma}u_k - \varphi^{\mathrm{T}}(\vec{u}_k)\tilde{\Gamma}\delta_k \le 0 \\
\Longleftrightarrow&\beta_k^{\mathrm{T}} \Pi_k \beta_k \le 0
\end{aligned} \tag{10.40}$$

where Π_k is defined in (10.25).

Likewise, we have

$$\beta_k^{\mathrm{T}} \Upsilon_{i,k} \beta_k \le 0 \tag{10.41}$$

where $\Upsilon_{i,k}$ is defined in (10.26).

For the brevity of presentation, we denote

$$\begin{aligned}
N_{i,k} &\triangleq \mathrm{diag}\left\{-1, \mathcal{T}_{q,i}^{\mathrm{T}} \mathcal{T}_{q,i}, 0, 0, 0, 0, 0, 0\right\}, \\
M_{1,k} &\triangleq \mathrm{diag}\left\{-1, 0, S_k^{-1}, 0, 0, 0, 0, 0\right\}, \\
J_{i,k} &\triangleq \mathrm{diag}\left\{-1, 0, 0, \mathcal{T}_{p,i}^{\mathrm{T}} R_{i,k}^{-1} \mathcal{T}_{p,i}, 0, 0, 0, 0\right\}, \\
M_{2,k} &\triangleq \mathrm{diag}\left\{-1, 0, 0, 0, W_k^{-1}, 0, 0, 0\right\}, \\
M_{3,k} &\triangleq \mathrm{diag}\left\{-1, 0, 0, 0, 0, V_k^{-1}, 0, 0\right\}.
\end{aligned}$$

According to Definition 10.2, it is not difficult to reformulate the constraints $\mathbb{E}\{z_{i,k}^{\mathrm{T}} z_{i,k}\} \le 1$, $\delta_k \in \mathscr{E}(0, S_k, m)$, $\vartheta_{i,k} \in \mathscr{E}(0, R_{i,k}, p)$, $w_k \in$

$\mathcal{E}(0, W_k, \omega)$ *and* $v_k \in \mathcal{E}(0, V_k, \nu)$ *in terms of the variable* β_k *as follows:*

$$\beta_k^{\mathrm{T}} N_{i,k} \beta_k \leq 0,$$
$$\beta_k^{\mathrm{T}} M_{1,k} \beta_k \leq 0,$$
$$\beta_k^{\mathrm{T}} J_{i,k} \beta_k \leq 0, \tag{10.42}$$
$$\beta_k^{\mathrm{T}} M_{2,k} \beta_k \leq 0,$$
$$\beta_k^{\mathrm{T}} M_{3,k} \beta_k \leq 0.$$

Having expressed the constraints uniformly in terms of a series of quadratic functions with regard to the variable β_k*, we are now going to demonstrate that the inequality* (10.32) *is true with the condition given in this theorem.*

By applying the Schur Complement Lemma (Lemma 10.1) to the set of RLMIs (10.23) *and noting that* $S_k = \tau_{1,k} \tilde{S}_k^{-1}$ *and* $R_{i,k} = \varrho_{i,k} \tilde{R}_{i,k}^{-1}$*, we obtain*

$$\bar{\Lambda}_k^{\mathrm{T}} \mathcal{T}_{n,i}^{\mathrm{T}} \Phi_{k+1}^{-1} \mathcal{T}_{n,i} \bar{\Lambda}_k + \hat{\Lambda}_k^{\mathrm{T}} \mathcal{T}_{n,i}^{\mathrm{T}} \Phi_{k+1}^{-1} \mathcal{T}_{n,i} \hat{\Lambda}_k - \Delta_k \leq 0 \tag{10.43}$$

which, by (10.24)*, is equivalent to*

$$\bar{\Lambda}_k^{\mathrm{T}} \mathcal{T}_{n,i}^{\mathrm{T}} \Phi_{k+1}^{-1} \mathcal{T}_{n,i} \bar{\Lambda}_k + \hat{\Lambda}_k^{\mathrm{T}} \mathcal{T}_{n,i}^{\mathrm{T}} \Phi_{k+1}^{-1} \mathcal{T}_{n,i} \hat{\Lambda}_k$$
$$-\mathrm{diag}\{1, 0, 0, 0, 0, 0, 0, 0\}$$
$$-\tau_{1,k} M_{1,k} - \tau_{2,k} M_{2,k} - \tau_{3,k} M_{3,k} - \tau_{4,k} \Pi_k$$
$$-\sum_{i=1}^{N} \epsilon_{i,k} N_{i,k} - \sum_{i=1}^{N} \rho_{i,k} \Upsilon_{i,k} - \sum_{i=1}^{N} \varrho_{i,k} J_{i,k} \leq 0. \tag{10.44}$$

By resorting to Lemma 10.2 and on account of (10.40)*,* (10.41) *and* (10.42)*, we immediately arrive at*

$$\beta_k^{\mathrm{T}} \left(\bar{\Lambda}_k^{\mathrm{T}} \mathcal{T}_{n,i}^{\mathrm{T}} \Phi_{k+1}^{-1} \mathcal{T}_{n,i} \bar{\Lambda}_k + \hat{\Lambda}_k^{\mathrm{T}} \mathcal{T}_{n,i}^{\mathrm{T}} \Phi_{k+1}^{-1} \mathcal{T}_{n,i} \hat{\Lambda}_k \right) \beta_k \leq 1, \tag{10.45}$$

which, by means of (10.39)*, further indicates that*

$$\mathbb{E}\{\beta_k^{\mathrm{T}} \Lambda_k^{\mathrm{T}} \mathcal{T}_{n,i}^{\mathrm{T}} \Phi_{k+1}^{-1} \mathcal{T}_{n,i} \Lambda_k \beta_k\} \leq 1, \tag{10.46}$$

or equivalently,

$$\mathbb{E}\{\tilde{x}_{i,k+1}^{\mathrm{T}} \Phi_{k+1}^{-1} \tilde{x}_{i,k+1}\} \leq 1. \tag{10.47}$$

By using the Schur Complement Lemma again, the inequality (10.47) *holds if and only if*

$$\mathbb{E}\{\tilde{x}_{i,k+1} \tilde{x}_{i,k+1}^{\mathrm{T}}\} \leq \Phi_{k+1}, \tag{10.48}$$

which indicates the accomplishment of the induction. Accordingly, we conclude that the design objective (10.17) *is achieved subject to the fixed communication topology* $\mathscr{G}$ *and variance constraints* $\{\Phi_k\}_{k \geq 0}$*. The desired sequences of filtering parameters* $\{\mathcal{G}_k\}_{k \geq 0}$ *and* $\{\mathcal{K}_k\}_{k \geq 0}$ *can be obtained by solving the set of RLMIs* (10.23) *iteratively. The proof is now complete.*

In the following stage, an optimization problem is formulated with the purpose of determining the filtering gains ensuring the locally optimal filtering performance by minimizing Φ_k in the sense of matrix trace at each time instant.

Corollary 10.1 *For the system* (10.1)–(10.2) *subject to the deception attacks* (10.6), *let the network topology $\mathcal{G}$ be given. A sequence of minimized $\{\Phi_k\}_{k\geq 1}$ can be guaranteed (in the sense of matrix trace) if there exist sequences of real-valued matrices $\{\mathcal{G}_k\}_{k\geq 0}$ and $\{\mathcal{K}_k\}_{k\geq 0}$ ($\mathcal{K}_k \in \mathcal{Q}_{n\times m}$), sequences of positive definite matrices $\{\tilde{S}_k\}_{k\geq 0}$ and $\{\tilde{R}_{i,k}\}_{k\geq 0}$, sequences of non-negative scalars $\{\tau_{1,k}\}_{k\geq 0}$, $\{\tau_{2,k}\}_{k\geq 0}$, $\{\tau_{3,k}\}_{k\geq 0}$, $\{\tau_{4,k}\}_{k\geq 0}$, $\{\epsilon_{i,k}\}_{k\geq 0}$, $\{\varrho_{i,k}\}_{k\geq 0}$ and $\{\rho_{i,k}\}_{k\geq 0}$ ($i = 1, 2, \ldots, N$) solving the following optimization problem:*

$$\min_{\Phi_{k+1},\mathcal{G}_k,\mathcal{K}_k,\tilde{S}_k,\tilde{R}_{i,k},\tau_{1,k},\tau_{2,k},\tau_{3,k},\tau_{4,k},\epsilon_{i,k},\varrho_{i,k},\rho_{i,k}} \operatorname{tr}[\Phi_{k+1}] \qquad (10.49)$$

$$\text{subject to} \quad \begin{bmatrix} -\Delta_k & * & * \\ \mathcal{T}_{n,i}\bar{\Lambda}_k & -\Phi_{k+1} & * \\ \mathcal{T}_{n,i}\hat{\Lambda}_k & 0 & -\Phi_{k+1} \end{bmatrix} \leq 0.$$

Remark 10.5 *So far, the addressed variance-constrained distributed filtering problem has been discussed for time-varying systems with multiplicative noises, unknown but bounded disturbances and deception attacks. The existence condition of the desired distributed filter has been established by resorting to the recursive linear matrix inequality approach. The filter gains can be determined via solving a set of RLMIs iteratively. The optimization problem for the locally optimal filtering performance has also been solved by minimizing the constraints imposed on the estimation error variance in the sense of matrix trace. One of our future research topics is the variance-constrained distributed filtering subject to other frequently seen network-induced complexities such as transmission delay [132] and quantization effects [97]. It is also our interest to apply the proposed technique to deal with consensus control problems for stochastic multi-agent systems studied in [141], where the open communication network may be also under attack.*

10.3   Numerical Example

In this section, an illustrative example is presented to illustrate the effectiveness of the proposed algorithm for the addressed variance-constrained distributed filtering problem for time-varying systems subject to multiplicative noises, unknown but bounded disturbances and deception attacks.

We consider a target tracking system whose dynamic model is given as follows:

$$x_{k+1} = \begin{bmatrix} 1 & T \\ 0 & 1 \end{bmatrix} + Dw_k \qquad (10.50)$$

where T is the sampling period and the state $x_k = [s_k \ \dot{s}_k]^{\mathrm{T}}$ with s_k and $\dot{s}_k$ being the position and velocity, respectively. It is worth mentioning in (10.50), the system parameters are time-invariant and the state is only subject to the external additive noise w_k. However, in real-world application, because of the changeable circumstance, these parameters are usually time-varying. Moreover, the system may contain, apart from additive noises, certain multiplicative disturbances that have significant impact on performance [237]. Taking these into account, we propose the model of system (10.1)–(10.2) with following parameters that are of more practical significance:

$$A_k = \begin{bmatrix} 1 & T \\ 0 & 1 \end{bmatrix}, \ T = 0.1,$$

$$A_{1,k} = \begin{bmatrix} 0.08 + 0.1\sin(3k) & -0.05 \\ 0.01 & -0.01 + 0.02\cos(k) \end{bmatrix},$$

$$A_{2,k} = \begin{bmatrix} 0.06 + 0.01\sin(10k) & 0.02 \\ -0.01 & -0.07 + 0.02\cos(k) \end{bmatrix},$$

$$C_{1,k} = \begin{bmatrix} 1 & 0 \end{bmatrix},$$

$$C_{2,k} = \begin{bmatrix} 0.9 + \sin(k) & 0 \end{bmatrix},$$

$$C_{3,k} = \begin{bmatrix} 0.75 + 0.5\sin(k) & 0.1 \end{bmatrix},$$

$$D_k = \begin{bmatrix} 0.1 + 0.05\cos(3k) \\ 0.2 + 0.04\exp\{-k\} \end{bmatrix},$$

$$E_{1,k} = 0.2 + 0.05\cos(3k),$$

$$E_{2,k} = 0.2 + 0.15\sin(2k),$$

$$E_{3,k} = 0.15 + 0.05\sin(2k).$$

Suppose the unknown but bounded disturbances are $w_k = 0.2\sin(5k)$ and $v_k = 0.5\cos(2k)$ and set $W_k = 0.04$, $V_k = 0.25$. Then it can be easily checked that w_k and v_k belong to the ellipsoid sets defined in (10.5).

Suppose that there are three sensor nodes connected according to graph $\mathscr{G}$. The associated adjacency matrix $\mathscr{L}$ is selected as follows:

$$\mathscr{L} = \begin{bmatrix} 1 & 0 & 1 \\ 0 & 1 & 1 \\ 1 & 1 & 1 \end{bmatrix}.$$

Set the initial values as follows:

$$x_0 = \begin{bmatrix} 5 \\ 3 \end{bmatrix}, \ \hat{x}_{1,0} = \begin{bmatrix} 2.8 \\ 1.6 \end{bmatrix}, \ \hat{x}_{2,0} = \begin{bmatrix} 3.0 \\ 2.0 \end{bmatrix},$$

$$\hat{x}_{3,0} = \begin{bmatrix} 3.2 \\ 1.2 \end{bmatrix}, \ \Phi_0 = \begin{bmatrix} 15 & 0.1 \\ 0.1 & 15 \end{bmatrix}.$$

Then it can be easily checked that the initial condition (10.16) is satisfied.

Suppose that the constraints imposed on the attack signals are characterized as follows:

$$\underline{\Gamma} = 0.8, \ \bar{\Gamma} = 1.2, \ \underline{\Xi}_1 = 0.8, \ \bar{\Xi}_1 = 1.2, \ \underline{\Xi}_2 = 0.9,$$

$$\bar{\Xi}_2 = 1.1, \ \underline{\Xi}_3 = 0.7, \ \bar{\Xi}_3 = 1.3.$$

In this section, we proceed to utilize the algorithm proposed in Corollary 10.1 to solve the optimization problem (10.49). By using Matlab software, the simulation is carried out and the results are shown in Figs. 10.2–10.7. Specifically, Figs. 10.2–10.3 depict the trajectories of the individual entries of

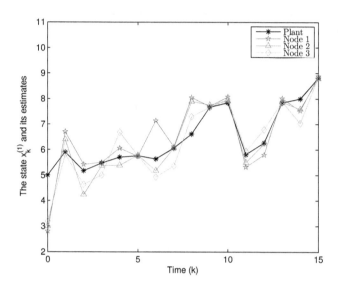

FIGURE 10.2: The state $x_k^{(1)}$ and its estimate.

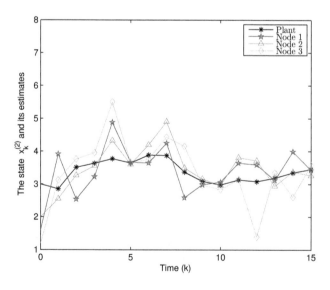

FIGURE 10.3: The state $x_k^{(2)}$ and its estimate.

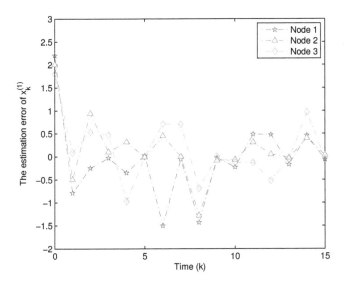

FIGURE 10.4: The estimation error of $x_k^{(1)}$.

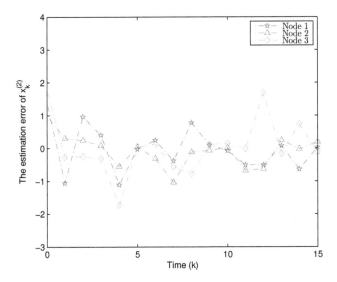

FIGURE 10.5: The estimation error of $x_k^{(2)}$.

the system state x_k (i.e., $x_k^{(1)}$ and $x_k^{(2)}$) and their estimates at each sensing node. Figs. 10.4–10.5 present the trajectories of the one-step-ahead estimation errors (i.e., $\tilde{x}_{i,k}$) of all the sensing nodes. The estimation error variances of

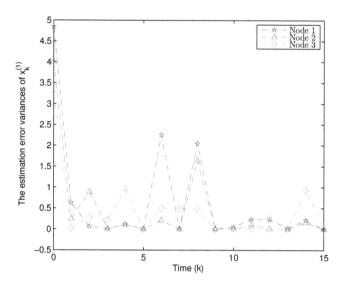

FIGURE 10.6: The variance of $\tilde{x}_k^{(1)}$.

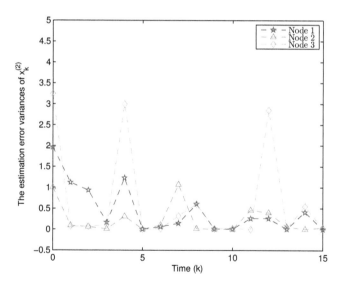

FIGURE 10.7: The variance of $\tilde{x}_k^{(2)}$.

all the sensing nodes are proposed in Fig. 10.6–10.7, which indicate that the proposed algorithm performs quite well.

10.4 Summary

In this chapter, the variance-constrained distributed filtering problem has been studied for a class of discrete time-varying systems with multiplicative noises, unknown but bounded disturbances and deception attacks over sensor networks. A novel deception attack model has been proposed, where the attack signals are injected by the adversary to both control and measurement data during the transmission via the communication networks. A sufficient condition has been established for the existence of the required filter satisfying the estimation error variance constraints by means of the RLMI approach. An optimization problem has been presented to seek the filter parameters with the guarantee of the locally minimal estimation error variance at each time instant. Finally, an illustrative example has been used to show the effectiveness and applicability of the proposed algorithm.

11

Conclusions and Future Topics

In this book, recent advances in control and filtering problems have been investigated for stochastic networked systems. Latest results on analysis and synthesis problems for systems in the context of several different sorts of networks first have been surveyed. Then, in each chapter, for the addressed stochastic networked systems, sufficient conditions have been derived for the closed-loop systems to achieve certain pre-specified desired performances including stability, robustness, consensus and accuracy. Subsequently, the controller/fitler synthesis problems have been investigated, and sufficient conditions for the existence of the desired controller/filter with which all the desired pre-specified system performance requirements can be satisfied have been given by means of certain convex optimization algorithms.

This book has established a unified theoretical framework for analysis and synthesis of control/filteirng problems in the context of several different sorts of stochastic networked control systems. However, the obtained results are still quite limited. Some of the possible future research topics are listed as follows:

In real-world engineering applications, there are still some more complicated yet important categories of stochastic multi-agent systems that have not attracted adequate research attention. As a consequence, the consensus control problems still remain open and challenging for MASs whose dynamics are more general and complex such as those investigated in [22, 45, 46, 122, 127, 136, 138, 140, 143, 249, 259]. Moreover, in practice, since the energy that each agent could use is always limited, another future research direction is to further investigate the consensus control problem for stochastic MASs while using the full/better utilization of the limited energy. The event-based communication mechanism [40,41,50,123,217,232,256] might be a possible way to solve the energy-constrained consensus control problem. Furthermore, since the communication protocol plays a paramount role in the consensus of multi-agent systems, more effort should be devoted to the investigation of the protocol design issues for MASs in the presence of stochasticity. Different protocols, such as those studied in [271, 273], could be adopted to enhance the consensus performance of stochastic multi-agent systems.

On the other hand, most of the existing literature regarding the addressed problems assumes that the topologies are fixed and/or time-invariant, and the limited work concerning the case of random topologies is mainly focused on the Markovian switching topology. However, in practical engineering, sometimes such a random topology could not reflect explicitly the real situation of the time-varying and stochastic topology. As a result, it would be interesting to study the consensus problem for MASs with topology varying according to other probabilistic distribution rather than a Markov chain.

Bibliography

[1] F. Abdollahi and K. Khorasani. A decentralized Markovian jump $\mathcal{H}_\infty$ control routing strategy for mobile multi-agent networked systems. *IEEE Transactions on Control Systems Technology*, 19(2):269–283, 2011.

[2] K. Abidi, J. Xu, and X. Yu. On the discrete-time integral sliding-mode control. *IEEE Transactions on Automatic Control*, 52(4):709–715, 2007.

[3] N. Amelina and A. Fradkov. Approximate consensus in multi-agent nonlinear stochastic systems. In *Proceedings of the European Control Conference (ECC), Strasbourg, France*, pages 2833–2838, 2014.

[4] S. Amin, G. A. Schwartz, A. A. Cardenas, and S. S. Sastry. Game-theoretic models of electricity theft detection in smart utility networks providing new capabilities with advanced metering infrastructure. *IEEE Control Systems Magazines*, 35(1):66–81, 2015.

[5] S. Amin, G. A. Schwartz, and S. S. Sastry. Security of interdependent and identical networked control systems. *Automatica*, 49(1):186–192, 2013.

[6] R. D. Andrea and G. E. Dullerud. Distributed control design for spatially interconnected systems. *IEEE Transactions on Automatic Control*, 48(9):1478–1495, 2003.

[7] K. Åström. *Introduction to Stochastic Control Theory*. Academic Press. New York and London, 1970.

[8] S. Azuma, I. Baba, and T. Sugie. Broadcast control of Markovian multi-agent systems. *SICE Journal of Control, Measurement, and System Integration*, 9(2):103–112, 2016.

[9] Y. Azuma and T. Ohtsuka. Receding horizon nash game approach for distributed nonlinear control. In *Proceedings of the SICE Annual Conference, Tokyo, Japan*, pages 380–384, 2011.

[10] Z. A. Baig and K. Salah. Multi-agent pattern recoginition mechanism for detecting distributed denial of service attacks. *IET Information Security*, 4(4):333–343, 2010.

[11] M. Basin, S. Elvira-Ceja, and E. Sanchez. Central suboptimal mean-square H_∞ controller design for linear stochastic time-varying systems. *International Journal of Systems Science*, 42(5):821–827, 2011.

[12] D. Bernstein and W. Haddad. Steady-state Kalman filtering with an H_∞ error bound. *Systems and Control Letters*, 12(1):9–16, 1989.

[13] D. P. Bertsekas and I. B. Rhodes. Recursive state estimation for a set-membership description of uncertainty. *IEEE Transactions on Automatic Control*, 16(2):117–128, 1971.

[14] S. Bijani and D. Robertson. A review of attacks and security approaches in open multi-agent systems. *Artificial Intelligence Review*, 42(4):607–636, 2014.

[15] M. Blanke, M. Kinnaert, J. Lunze, M. Staroswiecki, and J. Schröder. *Diagnosis and fault-tolerant control*. Springer-Verlag New York, 2006.

[16] C. Bonivento, M. Sandri, and R. Zanasi. Discrete variable structure integral controllers. *Automatica*, 34:355–361, 1998.

[17] M. Boukhnifer and A. Chaibet. Robust control of aerial vehicle flight: Simulation and experimental results. *IEEE Aerospace and Electronic Systems Magazine*, 29(9):4–12, 2014.

[18] S. Boyd, L.E. Ghaoui, E. Feron, and V. Balakrishnan. *Linear Matrix Inequalities in System and Control Theory*. SIAM Stud. Appl. Math., 1994.

[19] P. Braca, R. Goldhahn, G. Ferri, and K. D. LePage. Distributed information fusion in multistatic sensor networks for underwater surveillance. *IEEE Sensors Journal*, 16(11):4003–4014, 2016.

[20] V. Brinatti, D. Guilermo, O. Eduardo, and D. Kunik. Distributed temperature sensing using cyclic pseudorandom sequences. *IEEE Sensors Journal*, 17(6):1686–1691, 2017.

[21] R. Cabell, D. Palumbo, and J. Vipperman. A principal component feedforward algorithm for active noise control: Flight test results. *IEEE Transactions on Control System Technology*, 9(1):76–83, 2001.

[22] C. Cai, Z. Wang, J. Xu, X. Liu, and F. E. Alsaadi. An integrated approach to global synchronization and state estimation for nonlinear singularly perturbed complex networks. *IEEE Transactions on Cybernetics*, 45(8):1597–1609, 2015.

[23] A. A. Cárdenas, S. Amin, and S. S. Sastry. Secure control: Towards survivable cyber-physical systems. In *Proceedings of the 1st International Workshop on Cyber-Physical Systems, Beijing, China*, pages 495–500, 2008.

[24] C. Chan. Discrete adaptive sliding-mode control of a class of stochastic systems. *Automatica*, 35:1491–1498, 1999.

[25] K. Chang and W. Wang. Robust covariance control for perturbed stochastic multivariable system via variable structure control. *Systems and Control Letters*, 37:323–328, 1999.

[26] B. Chen, W. Chen, and H. Wu. Robust H_2/H_∞ global linearization filter design for nonlinear stochastic systems. *IEEE Transactions on Circuits and Systems–I: Regular Paper*, 56(7):1441–1454, 2009.

[27] B. Chen and W. Zhang. Stochastic H_2/H_∞ control with state-dependent noise. *IEEE Transactions on Automatic Control*, 49(1):45–57, 2004.

[28] B. Chen, W.-A. Zhang, and L. Yu. Distributed finite-horizon fusion Kalman filtering for bandwidth and energy constrained wireless sensor networks. *IEEE Transactions on Signal Processing*, 62(4):797–812, 2014.

[29] F. Chen, H. Yu, and X. Xia. Output consensus of multi-agent systems with delayed and sampled-data. *IET Control Theory and Applications*, 11(5):632–639, 2017.

[30] H. F. Chen. *Stochastic Approximation and its Applications*. Springer, 2002.

[31] J. Chen, D. Xie, and M. Yu. Consensus problem of networked multi-agent systems with constant communication delay: Stochastic switching topology case. *International Journal of Control*, 85(9):1248–1262, 2012.

[32] X. Chen and T. Fukuda. Robust adaptive quasi-sliding mode controller for discrete-time systems. *Systems and Control Letters*, 35:165–173, 1998.

[33] X. Chen and K. Zhou. Multi-objective H_2 and H_∞ control design. *SIMA Journal of Control optimization*, 40(2):628–660, 2001.

[34] C. Cheng, M. Lin, and J. Hsiao. Sliding mode controllers design for linear discrete-time systems with matching pertubations. *Automatica*, 36:1205–1211, 2000.

[35] Y. J. Cho and D. J. Park. Sequential detector design for spectrum sensing considering instantaneously nonidentically distributed samples. *IEEE Transactions on Vehicular Technology*, 66(3):2158–2169, 2017.

[36] M. Y. Chow and Y. Tipsuwan. Network-based control systems: a tutorial. In *Proceedings of IECON'01: the 27th Annual Conference of the IEEE Industrial Electronics Society, Denver, CO, USA*, pages 1593–1602, 2001.

[37] E. Collins and R. Skelton. A theory of state covariance assignment for discrete systems. *IEEE Transactions on Automatic Control*, 32(1):35–41, 1987.

[38] W. Cui, Y. Tang, J. A. Fang, and J. Kurth. Consensus analysis of second-order multi-agent networks with sampled data and packet losses. *IEEE Access*, 4:8127–8137, 2016.

[39] D. V. Dimarogonas, E. Frazzoli, and K. H. Johansson. Distributed event-triggered control for multi-agent systems. *IEEE Transactions on Automatic Control*, 57(5):1291–1297, 2012.

[40] D. Ding, Z. Wang, and B. Shen. Recent advances on distributed filtering for stochastic systems over sensor networks. *International Journal of General Systems*, 43(3–4):372–386, 2014.

[41] D. Ding, Z. Wang, B. Shen, and H. Dong. Event-triggered distributed H_∞ state estimation with packet dropouts through sensor networks. *IET Control Theory and Applications*, 9(13):1948–1955, 2015.

[42] D. Ding, Z. Wang, B. Shen, and G. Wei. Event-triggered consensus control for discrete-time stochastic multi-agent systems: the input-to-state stability in probability. *Automatica*, 62:284–291, 2015.

[43] L. Ding and G. Guo. Sampled-data leader-following consensus for nonlinear multi-agent systems with Markovian switching topologies and communication delay. *Journal of the Franklin Institute-Engineering and Applied Mathematics*, 352(1):369–383, 2015.

[44] H. Dong, Z. Wang, X. Bu, and F. E. Alsaadi. Distributed fault estimation with randomly occurring uncertainties over sensor networks. *International Journal of General Systems*, 45(5):662–674, 2016.

[45] H. Dong, Z. Wang, S. X. Ding, and H. Gao. Event-based H_∞ filter design for a class of nonlinear time-varying systems with fading channels and multiplicative noises. *IEEE Transactions on Signal Processing*, 63(13):3387–3395, 2015.

[46] H. Dong, Z. Wang, S. X. Ding, and H. Gao. Finite-horizon reliable control with randomly occurring uncertainties and nonlinearities subject to output quantization. *Automatica*, 52:355–362, 2015.

[47] H. Dong, Z. Wang, S. X. Ding, and H. Gao. On H_∞ estimation of randomly occurring faults for a class of nonlinear time-varying systems with fading channels. *IEEE Transactions on Automatic Control*, 61(2):479–484, 2016.

[48] H. Dong, Z. Wang, and H. Gao. Fault detection for Markovian jump systems with sensor saturations and randomly varying nonlinearities. *IEEE Transactions on Circuits and Systems - Part I*, 59(10):2354–2362, 2012.

[49] H. Dong, Z. Wang, and H. Gao. Distributed H_∞ filtering for a class of Markovian jump nonlinear time-delay systems over lossy sensor networks. *IEEE Transactions on Industrial Electronics*, 60(10):4665–4672, 2013.

[50] H. Dong, Z. Wang, J. Lam, and H. Gao. Distributed filtering in sensor networks with randomly occurring saturations and successive packet dropouts. *International Journal of Robust and Nonlinear Control*, 24(12):1743–1759, 2014.

[51] W. B. Dunbar. Distributed receding horizon control of dynamically coupled nonlinear systems. *IEEE Transactions on Automatic Control*, 52(7):1249–1263, 2007.

[52] N. Laiand, C. Edwards and S. Spurgeon. Discrete output feedback sliding-mode control with integral action. *International Journal of Robust Nonlinear Control*, 16:21–43, 2006.

[53] R. Eustace, B. Woodyatt, G. Merrington, and A. Runacres. Fault signatures obtained from fault implant tests on an F404 engine. *ASME Transactions, Journal of Engineering for Gas Turbines and Power*, 116(1):178–183, 1994.

[54] M. Fanaswala and V. Krishnamurthy. Detection of anomalous trajectory patterns in target tracking via stochastic context-free grammars and reciprocal process models. *IEEE Journal of Selected Topics in Signal Processing*, 7(1):76–90, 2013.

[55] L. Fang, P. J. Antsaklis, and A. Tzimas. Asynchronous consensus protocols: Preliminary results, simulations and open questions. In *Proceedings of the 44th IEEE Conference on Decision and Control, and the European Control Conference 2005, Seville, Spain*, pages 2194–2199, 2005.

[56] H. Fawzi, P. Tabuada, and S. Diggavi. Secure estimation and control for cyber-physical systems under adversarial attacks. *IEEE Transactions on Automatic control*, 59(6):1454–1467, 2014.

[57] J. A. Fax and R. M. Murray. Information flow and cooperative control of vehicle formations. *IEEE Transactions on Automatic Control*, 49(9):1465–1476, 2004.

[58] Z. Feng and G. Hu. Passivity-based consensus and passification for a class of stochastic multi-agent systems with switching topology. In *Proceedings of the 12th International Conference on Control, Automation, Robotics and Vision, Guangzhou, China*, pages 1460–1465, 2012.

[59] Z. Feng, G. Hu, and G. Wen. Distributed consensus tracking for multi-agent systems under two types of attacks. *International Journal of Robust and Nonlinear Control*, 26(5):896–918, 2016.

[60] Y. K. Foo, Y.C. Soh, and M. Moayedi. Linear set-membership state estimation with unknown but bounded disturbances. *International Journal of Systems Sciences*, 43(4):715–730, 2012.

[61] M. Franceschelli, A. Giua, and C. Seatzu. Distributed averaging in sensor networks based on broadcast gossip algorithms. *IEEE Sensors Journal*, 11(3):808–817, 2011.

[62] E. Franco, L. Magni, T. Parisini, and M. M. Polycarpou. Cooperative constrained control of distributed agents with nonlinear dynamics and delayed information exchange: A stabilizing receding-horizon approach. *IEEE Transactions on Automatic Control*, 53(1):324–338, 2008.

[63] H. Gao, J. Lam, L. Xie, and K. Chen. H_∞ fuzzy filtering of nonlinear systems with intermittent measurements. *IEEE Transactions on Fuzzy Systems*, 17(2):291–300, 2009.

[64] X. Ge and Q-L. Han. Distributed event-triggered H_∞ filtering over sensor networks with communication delays. *Information Science*, 291:128–142, 2015.

[65] E. Gershon, A. Pila, and U. Shaked. Difference LMIs for robust H_∞ control and filtering. In *Proceedings of the European Control Conference, Porto, Portugal*, pages 3469–3474, 2001.

[66] L. El. Ghaoui and G. Calafiore. Robust filtering for discrete-time systems with bounded noise and parametric uncertainty. *IEEE Transactions on Automatic Control*, 46:11084–1089, 2001.

[67] A. Giani, S. Sastry, K. H. Johansson, and H. Sandberg. The VIKING project: An initiative on resilient control of power networks. In *Proceedings of the 2nd International Symposium on Resilient Control Systems*, Idaho Falls, Idaho, 2009.

[68] D. Gillies, D. Thornley, and C. Bisdikian. Probabilistic approaches to estimating the quality of information in military sensor networks. *Computer Journal*, 53(5):493–502, 2010.

[69] G. Golo and Č. Milosavljević. Robust discrete-time chattering free sliding mode control. *Systems and Control Letters*, 41(0):19–28, 2000.

[70] D. P. Goodall and R. Postoyan. Output feedback stabilization for uncertain nonlinear time-delay systems subject to input constraints. *International Journal of Control*, 83(4):676–693, 2010.

[71] G. Guo, L. Ding, and Q.-L. Han. A distributed event-triggered transmission strategy for sampled-data consensus of multi-agent systems. *Automatica*, 50(5):1489–1496, 2014.

[72] L. Guo and Y-F. Huang. Frequency-domain set-membership filtering and its applications. *IEEE Transactions on Signal Processing*, 55(4):1326–1338, 2007.

[73] V. Gupta, B. Hassibi, and R. M. Murray. A sub-optimal algorithm to synthesize control laws for a network of dynamic agents. *International Journal of Control*, 78(16):1302–1313, 2005.

[74] M. Sznaier, H. Rotstein and M. Idan. Mixed H_2/H_∞ filtering theory and an aerospace application. *International Journal of Robust and Nonlinear Control*, 6:347–366, 1996.

[75] A. T. Hafez, A. J. Marasco, S. N. Givigi, M. Iskandarani, S. Yousefi, and C. A. Rabbath. Solving multi-UAV dynamic encirclement via model predictive control. *IEEE Transactions on Control Systems Technology*, 23(6):2251–2265, 2015.

[76] Y. Hatano and M. Mesbahi. Agreement over random networks. *IEEE Transactions on Automatic Control*, 50(11):1867–1872, 2005.

[77] D. He, C. He, L.-G. Jiang, H.-W. Zhu, and G.-R. Hu. Chaotic characteristics of a one-dimensional iterative map with infinite collapses. *IEEE Transactions on Circuits Systems*, 48(7):900–C906, 2001.

[78] W. He, B. Zhang, Q-L. Han, F. Qian, J. Kurth, and J. Cao. Leaderfollowing consensus of nonlinear multi-agent systems with stochastic sampling. *IEEE Transactions on Cybernetics*, 47(2):327–338, 2017.

[79] D.W.C. Ho and Y. Niu. Robust fuzzy design for nonlinear uncertain stochastic systems via sliding-mode control. *IEEE Transactions on Fuzzy Systems*, 15(3):350–358, 2007.

[80] B. Hou, F. Sun, H. Li, and G. Liu. Stationary and dynamic consensus of second-order multi-agent systems with Markov jumping input delays. *IET Control Theory and Applications*, 8(17):1905–1913, 2014.

[81] N. Hou, H. Dong, Z. Wang, W. Ren, and F. E. Alsaadi. Non-fragile state estimation for discrete Markovian jumping neural networks. *Neurocomputing*, 179:238–245, 2016.

[82] C. T. Hsueh, C. Y. Wen, and Y. C. Ouyang. A secure scheme against power exhausting attacks in hierarchical wireless sensor networks. *IEEE Sensors Journal*, 15(6):2590–3602, 2015.

[83] A. Hu, J. Cao, M. Hu, and L. Guo. Event-triggered consensus of Markovian jumping multi-agent systems with stochastic sampling. *IET Control Theory and Applications*, 9(13):1964–1972, 2015.

[84] J. Hu, Z. Wang, H. Gao, and L. Stergioulas. Probability-guaranteed H_∞ finite-horizon filtering for a class of nonlinear time-varying systems with sensor saturations. *Systems and Control Letters*, 61(4):477–484, 2012.

[85] J. Hu, Z. Wang, S. Liu, and H. Gao. A variance-constrained approach to recursive state estimation for time-varying complex networks with missing measurements. *Automatica*, 64:155–162, 2016.

[86] Y. Hu, J. Lam, and J. Liang. Consensus control of multi-agent systems with missing data in actuators and Markovian communication failure. *International Journal of Systems Science*, 44(10):1867–1878, 2013.

[87] B. Huang. Minimum variance control and performance assessment of time-variant processes. *Journal of Process Control*, 12(6):707–719, 2002.

[88] M. Huang and J. Manton. Coordination and consensus of networked agents with noisy measurements: Stochastic algorithms and asymptotic behavior. *SIAM Journal on Control and Optimization*, 48(1):134–161, 2009.

[89] S. Hui and S.H. Żak. On discrete-time variable structure sliding mode control. *Systems and Control Letters*, 38:283–288, 1999.

[90] Y. Hung and F. Yang. Robust H_∞ filtering with error variance constraints for uncertain discrete time-varying systems with uncertainty. *Automatica*, 39(7):1185–1194, 2003.

[91] C. Hwang. Robust discrete variable structure control with finite-time approach to switching surface. *Automatica*, 38:167–175, 2002.

[92] J. Kim, J. Yang and H. Shim. Practical consensus for heterogeneous linear time-varying multi-agent systems. In *Proceedings of the 12th International Conference on Control, Automation and Systems, Korea*, pages 23–28, 2012.

[93] A. Jadbabaie, N. Motee, and M. Barahona. On the stability of the Kuramoto model of coupled nonlinear oscillators. In *Proceedings of the 2004 American Control Conference, Boston, Massachusetts, USA*, pages 4296–4301, 2004.

[94] R. Kalman and R. Bucy. A new approach to linear filtering and prediction problems. *ASME Journal of Basic Engineering*, 82:34–45, 1960.

[95] R. Kalman and R. Bucy. New results in linear filtering and prediction theory. *ASME Journal of Basic Engineering*, 83:95–107, 1961.

[96] S. Kar and J. M. F. Moura. Distributed consensus algorithms in sensor networks with imperfect communication: Link failures and channel noise. *IEEE Transactions on Signal Processing*, 57(1):355–369, 2009.

[97] H. R. Karimi. Robust H_∞ filter design for uncertain linear systems over network with network-induced delays and output quantization. *Model, Identification, and Control*, 30(1):27–37, 2009.

[98] H. R. Karimi. Robust delay-dependent H_∞ control of uncertain time-delay systems with mixed neutral, discrete, and distributed time-delays and Markovian switching parameters. *IEEE Transactions on Circuits and Systems I-Regular Papers*, 58(8):1910–1923, 2011.

[99] H. R. Karimi. A sliding mode approach to H_∞ synchronization of master-slave time-delay systems with Markovian jumping parameters and nonlinear uncertainties. *Journal of the Franklin Institute-Engineering and Applied Mathematics*, 349(4):1480–1496, 2012.

[100] H. R. Karimi. Passivity-based output feedback control of Markovian jump systems with discrete and distributed time-varying delays. *International Journal of Systems Science*, 44(7):1290–1300, 2013.

[101] H. R. Karimi, M. Zapateiro, and N. Luo. An LMI approach to vibration control of base-isolated building structures with delayed measurements. *International Journal of Systems Science*, 41(12):1511–1523, 2010.

[102] H. Katayama. Design of consensus controllers for multi-rate sampled-data strict-feedback multi-agent systems. *IFAC Papers Online*, 48(18):157–163, 2015.

[103] H. K. Khalil. *Nonlinear Systems*. Upper Saddle River, Prentice-Hall, NJ, 1996.

[104] N. Kita and A. Shimomura. Asymptotic behavior of solutions to Schrödinger equations with a subcritical dissipative nonlinearity. *Journal of Differential Equations*, 242(1):192–210, 2007.

[105] A. Komaee and P. I. Barton. Potential canals for control of nonlinear stochastic systems in the absence of state measurements. *IEEE Transactions on Control Systems Technology*, 25(1):161–174, 2017.

[106] M. Kumar, N. Stoll, R. Stoll, and K. Thurow. A stochastic framework for robust fuzzy filtering and analysis of signals - part II. *IEEE Transactions on Cybernetics*, 45(3):486–496, 2015.

[107] A. Kurzhanski and I. Vályi. *Ellipsoidal calculus for estimation and control*. Boston, MA: Birkhäuser, 1997.

[108] H. J. Kushner and G. Yin. *Stochastic Approximation and Recursive Algorithms and Applications*. Springer, 2003.

[109] N. Lai, C. Edwards, and S. Spurgeon. On output tracking using dynamic output feedback discrete-time sliding-mode controllers. *IEEE Transactions on Automatic Control*, 52(10):1975–1981, 2007.

[110] C. Langbort, R. S. Chandra, and R. D. Andrea. Distributed control design for systems interconnected over an arbitrary graph. *IEEE Transactions on Automatic Control*, 49(9):1502–1519, 2004.

[111] Q. Li, B. Shen, Y. Liu, , and F. E. Alsaadi. Event-triggered H_∞ state estimation for discrete-time stochastic genetic regulatory networks with Markovian jumping parameters and time-varying delays. *Neurocomputing*, 174:912–920, 2016.

[112] T. Li and J. F. Zhang. Mean square average consensus under measurement noises and fixed topologies: Necessary and sufficient conditions. *Automatica*, 45(8):1929–1936, 2009.

[113] T. Li and J. F. Zhang. Consensus conditions of multi-agent systems with time-varying topologies and stochastic communication noises. *IEEE Transactions on Automatic Control*, 55(9):2043–2057, 2010.

[114] W. Li, H. Zhou, Z. W. Liu, Y. Qin, and Z. Wang. Impulsive coordination of nonlinear multi-agent systems with multiple leaders and stochastic disturbance. *Neurocomputing*, 171:73–81, 2016.

[115] Z. Li, Z. Duan, and F. L. Lewis. Distributed robust consensus control of multi-agent systems with heterogeneous matching uncertainties. *Automatica*, 50(3):883–889, 2014.

[116] Q. Ling and M. Lemmon. Optimal dropout compensation in networked control systems. In *Proceedings of the IEEE Conference on Decision and Control, Maui, HI, USA*, pages 670–675, 2003.

[117] H. Liu, Z. Wang, B. Shen, and F. E. Alsaadi. State estimation for discrete-time memristive recurrent neural networks with stochastic time-delays. *International Journal of General Systems*, 45(5):633–647, 2016.

[118] J. Liu, X. Liu, W. C. Xie, and H. Zhang. Stochastic consensus seeking with communication delays. *Automatica*, 47(12):2689–2696, 2011.

[119] J. Liu, P. Ming, and S. Li. Consensus gain conditions of stochastic multi-agent system with communication noise. *International Journal of Control, Automation and Systems*, 14(5):1223–1230, 2016.

[120] J. Liu, H. Zhang, X. Liu, and W. C. Xie. Distributed stochastic consensus of multi-agent systems with noise and delayed measurements. *IET Control Theory and Applications*, 7(10):1359–1369, 2013.

[121] K. Liu, H. Zhu, and J. Lu. Bridging the gap between transmission noise and sampled data for robust consensus of multi-agent systems. *IEEE Transactions on Circuits and Systems I–Regular Papers*, 62(7):1836–1844, 2015.

[122] Q. Liu, Z. Wang, X. He, G. Ghinea, and F. E. Alsaadi. A resilient approach to distributed filter design for time-varying systems under stochastic nonlinearities and sensor degradation. *IEEE Transactions on Signal Processing*, 65(5):1300–1309, 2017.

[123] Q. Liu, Z. Wang, X. He, and D. H. Zhou. Event-based H_∞ consensus control of multi-agent systems with relative output feedback: The finite-horizon case. *IEEE Transactions on Automatic Control*, 60(9):2553–2558, 2015.

[124] S. Liu. The Dirichlet problem with sublinear indefinite nonlinearities. *Nonlinear Analysis*, 73(9):2831–2841, 2010.

[125] Y. Liu, F. E. Alsaadi, X. Yin, and Y. Wang. Robust H_∞ filtering for discrete nonlinear delayed stochastic systems with missing measurements and randomly occurring nonlinearities. *International Journal of General Systems*, 44(2):169–181, 2015.

[126] Y. Liu, W. Liu, M. A. Obaid, and I. A. Abbas. Exponential stability of Markovian jumping Cohen-Grossberg neural networks with mixed mode-dependent time-delays. *Neurocomputing*, 177:409–415, 2016.

[127] Y. Liu, Z. Wang, X. He, and D. H. Zhou. H_∞ filtering for nonlinear systems with stochastic sensor saturations and Markov time-delays: The asymptotic stability in probability. *IET Control Theory and Applications*, 10(14):1706–1715, 2016.

[128] Y. Liu, Z. Wang, and W. Wang. Reliable $\mathcal{H}_\infty$ filtering for discrete time-delay systems with randomly occurred nonlinearities via delay-partitioning method. *Signal Processing*, 91:713–727, 2011.

[129] M. Long, C. H. Wu, and J. Y. Hung. Denial of service attacks on network-based control systems: impact and mitigation. *IEEE Transactions on Industrial Informatics*, 1(2):85–96, 2005.

[130] X. Lu, L. Xie, H. Zhang, and W. Wang. Robust Kalman filtering for discrete-time systems with measurement delay. *IEEE Transactions on Circuits and Systems-II: Express Briefs*, 54(6):522–526, 2007.

[131] Z. Lu and L.vS. Shieh. Simplex sliding mode control for nonlinear uncertain systems via chaos optimization. *Chaos, Solitions and Fractals*, 23:747–755, 2005.

[132] Y. Luo, G. Wei, Y. Liu, and X. Ding. Reliable H_∞ state estimation for 2-D discrete systems with infinite distributed delays and incomplete observations. *International Journal of General Systems*, 44(2):155–168, 2015.

[133] C. Q. Ma and Z.-Y. Qin. Bipartite consensus on networks of agents with antagonistic interactions and measurement noises. *IET Control Theory and Applications*, 10(17):2306–2313, 2016.

[134] L. Ma, Z. Wang, and Y. Bo. *Variance-constrained Multi-objective Stochastic control and filtering*. John Wiley and Sons, 2015.

[135] L. Ma, Z. Wang, Y. Bo, and Z. Guo. Robust H_∞ control of time-varying systems with stochastic non-linearities: the finite-horizon case. *Proceedings of the Institution of Mechanical Engineers, Part I (Journal of Systems and Control Engineering)*, 224:575–585, 2010.

[136] L. Ma, Z. Wang, Y. Bo, and Z. Guo. A game theory approach to mixed H_2/H_∞ control for a class of stochastic time-varying systems with randomly occurring nonlinearities. *Systems and Control Letters*, 60(12):1009–1015, 2011.

[137] L. Ma, Z. Wang, Y. Bo, and Z. Guo. Finite-horizon H_2/H_∞ control for a class of nonlinear Markovian jump systems with probabilistic sensor failures. *International Journal of Control*, 84(11):1847–1857, 2011.

[138] L. Ma, Z. Wang, Y. Bo, and Z. Guo. Robust H_∞ sliding mode control for nonlinear stochastic systems with multiple data packet losses. *International Journal of Robust and Nonlinear Control*, 22(5):473–491, 2012.

[139] L. Ma, Z. Wang, and Z. Guo. Robust fault-tolerant control for a class of nonlinear stochastic systems with variance constraints. *Journal of Dynamic Systems, Measurement, and Control*, 132(Article ID 044501), 2010.

[140] L. Ma, Z. Wang, Q. L. Han, and H. K. Lam. Variance-constrained distributed filtering for time-varying systems with multiplicative noises and deception attacks over sensor networks. *IEEE Sensors Journal*, 17(7):2279–2288, 2017.

[141] L. Ma, Z. Wang, and H. K. Lam. Event-triggered mean-square consensus control for time-varying stochastic multi-agent system with sensor saturations. *IEEE Transactions on Automatic Control*, 62(7):3524–3531, 2017.

[142] L. Ma, Z. Wang, and H. K. Lam. Mean-square H_∞ consensus control for a class of nonlinear time-varying stochastic multiagent systems: the finite-horizon case. *IEEE Transactions on Systems, Man, and Cybernetics: Systems*, 47(7):1050–1060, 2017.

[143] L. Ma, Z. Wang, H. K. Lam, F. E. Alsaadi, and X. Liu. Robust filtering for a class of nonlinear stochastic systems with probability constraints (invited paper). *Automation and Remote Control*, 77(1):37–54, 2016.

[144] L. Ma, Z. Wang, H. K. Lam, and N. Kyriakoulis. Distributed event-based set-membership filtering for a class of nonlinear systems with sensor saturations over sensor networks. *IEEE Transactions on Cybernetics*, 47(11):3772–3783, 2017.

[145] D. Mascarenas, E. Flynn, C. Farrar, G. Park, and M. Todd. A mobile host approach for wireless powering and interrogation of structural health monitoring sensor networks. *IEEE Sensors Journal*, 9(12):1719–1726, 2009.

[146] K. Mathiyalagan, P. Shi, H. Su, and R. Sakthivel. Exponential H_∞ filtering for discrete-time switched neural networks with random delays. *IEEE Transactions on Cybernetics*, 45(4):676–687, 2015.

[147] A. Mesbah. Stochastic model predictive control: An overview and prespective for future research. *IEEE Control Systems Magazine*, 36(6):30–44, 2016.

[148] N. Meskin and K. Khorasani. Fault Detection and Isolation of discrete-time Markovian jump linear systems with application to a network of multi-agent systems having imperfect communication channels. *Automatica*, 45(9):2032–2040, 2009.

[149] G. Miao and T. Li. Mean square containment control problems of multi-agent systems under Markov switching topologies. *Advances in Difference Equations*, 2015(Article ID 157), 2015.

[150] G. Miao, S. Xu, B. Zhang, and Y. Zou. Mean square consensus of second-order multi-agent systems under Markov switching topologies. *IMA Journal of Mathematical Control and Information*, 31(2):151–164, 2014.

[151] M. Milanese and A. Vicino. Optimal estimation theory for dynamic systems with set membership uncertainty: an overview. *Automatica*, 27:997–1009, 1991.

[152] P. Ming, J. Liu, S. Tan, S. Li, L. Shang, and X. Yu. Consensus stabilization in stochastic multi-agent systems with Markovian switching topology, noises and delay. *Neurocomputing*, 200:1–10, 2016.

[153] P. Ming, J. Liu, S. Tan, G. Wang, L. Shang, and C. Jia. Optimal estimation theory for dynamic systems with set membership uncertainty: an overview. *Journal of the Franklin Institute-Engineering and Applied Mathematics*, 352(9):3684–3700, 2015.

[154] S. Mini, S. K. Udgata, and S. L. Sabat. Sensor deployment and scheduling for target coverage problem in wireless sensor networks. *IEEE Sensors Journal*, 14(3):636–644, 2014.

[155] L. Mo and T. Pan. Mean-square consensus of heterogeneous multi-agent systems under Markov switching topologies. *Scientia Sinica Informationis*, 46(11):1621–1632, 2016.

[156] Y. Mo, S. Weerakkody, and B. Sinopoli. Physical authentication of control systems designing watermarked control inputs to detect counterfeit sensor outputs. *IEEE Control Systems Magazine*, 35(1):93–109, 2015.

[157] R. Morita, T. Wada, I. Masubuchi, T. Asai, and Y. Fujisaki. Multiagent consensus with noisy communication: Stopping rules based on network graphs. *IEEE Transactions on Control of Network Systems*, 3(4):358–365, 2016.

[158] R. Morita, T. Wada, I. Masubuchi, T. Asai, and Y. Fujisaki. Time averaging algorithms with stopping rules for multi-agent consensus with noisy measurements. *Asian Journal of Control*, 18(6):1969–1982, 2016.

[159] X. Mu and B. Zheng. Containment control of second-order discrete-time multi-agent systems with Markovian missing data. *IET Control Theory and Applications*, 9(8):1229–1237, 2015.

[160] X. Mu, B. Zheng, and K. Liu. $L_2 - L_\infty$ containment control of multi-agent systems with Markovian switching topologies and non-uniform time-varying delays. *IET Control Theory and Applications*, 8(10):863–872, 2014.

[161] Y. Nesterov and A. Nemirovski. *Interior point polynomial methods in convex programming: Theory and applications.* Philadelphia, PA: SIAM, 1994.

[162] W. Ni, X. Wang, and C. Xiong. Leader-following consensus of multiple linear systems under switching topologies: An averaging method. *Kybernetika*, 48(6):1194–1210, 2012.

[163] W. Ni, X. Wang, and C. Xiong. Consensus controllability, observability and robust design for leader-following linear multi-agent systems. *Automatica*, 49(7):2199–2205, 2013.

[164] W. Ni, D. Zhao, Y. Ni, and X. Wang. Stochastic averaging approach to leader-following consensus of linear multi-agent systems. *Journal of the Franklin Institute*, 353:2650–2669, 2016.

[165] J. Nilsson, B. Bernhardsson, and B. Wittenmark. Stochastic analysis and control of real-time systems with random time delays. *Automatica*, 34:57–64, 1998.

[166] Y. Niu and D.W.C. Ho. Robust observer design for Itô stochastic time-delay systems via sliding mode control. *Systems and Control Letters*, 55:781–793, 2006.

[167] Y. Niu, D.W.C. Ho, and J. Lam. Robust integral sliding mode control for uncertain stochastic systems with time-varying delay. *Automatica*, 41:873–880, 2005.

[168] B. Øksendal. *Stochastic Differential Equations: An Introduction with Applications*. Springer: Berlin, 2000.

[169] R. Olfati-Saber. Ultrafast consensus in small-world networks. In *Proceedings of 2005 American Control Conference, Portland, OR, USA*, pages 2372–2378, 2005.

[170] R. Olfati-Saber. Flocking for multi-agent dynamic systems: Algorithms and theory. *IEEE Transactions on Automatic Control*, 51(3):401–420, 2006.

[171] R. Olfati-Saber. Distributed Kalman filtering for sensor networks. *Proceedings of 46th IEEE Conference on Decision and Control*, 1–14:1814–1820, 2007.

[172] R. Olfati-Saber and J. S. Shamma. Consensus filters for sensor networks and distributed sensor fusion. In *Proceedings of 44th IEEE Conference on Decision and Control, and the European Control Conference 2005, Seville, Spain*, pages 6698–6703, 2005.

[173] Z. H. Pang and G. P. Liu. Design and implementation of secure networked predictive control systems under deception attacks. *IEEE Transactions on Control Systems Technology*, 20(5):1334–1342, 2012.

[174] M. J. Park, O. M. Kwon, J. H. Park, S. M. Lee, and E. J. Cha. Randomly changing leader-following consensus control for Markovian switching multi-agent systems with interval time-varying delays. *Nonlinear Analysis-Hybrid Systems*, 12:117–131, 2014.

[175] Y. Pei and J. Sun. Consensus of discrete-time linear multi-agent systems with Markov switching topologies and time-delay. *Neurocomputing*, 151:776–781, 2015.

[176] A. S. Poznyak. Sliding mode control in stochastic continuous-time systems: μ-zone MS-convergence. *IEEE Transactions on Automatic Control*, 62(2):863–868, 2017.

[177] I. Rapoport and Y. Oshman. Weiss-Weinstein lower bounds for Markovian systems. Part 2: Applications to fault-tolerant filtering. *IEEE Transactions on Signal Processing*, 55(5):2031–2042, 2007.

[178] W. Ren, K. Moore, and Y. Chen. High-order consensus algorithms in cooperative vehicle systems. In *Proceedings of the IEEE International Conference on Networking, Sensing and Control, Fort Lauderdale, USA*, pages 457–462, 2006.

[179] H. Robbins and S. Monro. A stochastic approximation method. *The Annals of Mathematical Statistics*, 22(3):400–407, 1951.

[180] A. Sahoo and S. Jagannathan. Stochastic optimal regulation of nonlinear networked control systems by using event-driven adaptive dynamic programming. *IEEE Transactions on Cybernetics*, 47(2):425–438, 2017.

[181] R. Sakthivel, S. Mohanapriya, H. R. Karimi, and P. Selvaraj. A robust repetitive-control design for a class of uncertain stochastic dynamical systems. *IEEE Transactions on Circuits and Systems II: Express Briefs*, 64(4):427–431, 2017.

[182] S. Satoh, H. J. Kappen, and M. Saeki. An iterative method for nonlinear stochastic optimal control based on path integrals. *IEEE Transactions on Automatic Control*, 62(1):262–276, 2017.

[183] A. V. Savkin and I. R. Petersen. Robust state estimation and model validation for discrete-time uncertain systems with a deterministic description of noise and uncertainty. *Automatica*, 34(2):271–274, 1998.

[184] L. Schenato and F. Firrentin. Average TimeSynch: A consensus-based protocol for clock synchronization in wireless sensor networks. *Automatica*, 47(9):1878–1886, 2011.

[185] C. Scherer. Multi-objective H_2/H_∞ control. *IEEE Transactions on Automatic Control*, 40(6):1054–1062, 1995.

[186] B. Schnaufer and W. Jenkins. Adaptive fault tolerance for reliable LMS adaptive filtering. *IEEE Transactions on Circuits and Systems - II: Analog and Digital Signal Processing*, 44(4):1001–1014, 1997.

[187] P. Seiler and R. Sengupta. An H_∞ approach to networked control. *IEEE Transactions on Automatic Control*, 50(3):356–364, 2005.

[188] L. Shang and X. Wang. Quickest attack detection in multi-agent reputation systems. *IEEE Journal of Selected Topics in Signal Processing*, 8(4):653–666, 2014.

[189] Y. Shang. Consensus recovery from intentional attacks in directed nonlinear multi-agent systems. *International Journal of Nonlinear Sciences and Numerical Simulation*, 14(6):355–361, 2013.

[190] Y. Shang. Couple-group consensus of continuous-time multi-agent systems under Markovian switching topologies. *Journal of the Franklin Institute-Engineering and Applied Mathematics*, 352(11):4826–4844, 2015.

[191] B. Shen, Z. Wang, and X. Liu. A stochastic sampled-data approach to distributed H_∞ filtering in sensor networks. *IEEE Transactions on Circuits and Systems - Part I*, 58(9):2237–2246, 2011.

[192] B. Shen, Z. Wang, and X. Liu. Sampled-data synchronization control of complex dynamical networks with stochastic sampling. *IEEE Transactions on Automatic Control*, 57(10):2644–2650, 2012.

[193] B. Shen, Z. Wang, H. Shu, and G. Wei. Robust $\mathcal{H}_\infty$ finite-horizon filtering with randomly occurred nonlinearities and quantization effects. *Automatica*, 46(11):1743–1751, 2010.

[194] B. Sinopoli, C. Sharp, L. Schenato, S. Schaffert, and S. S. Sastry. Distributed control applications within sensor networks. *Proceedings of the IEEE*, 91(8):1235–1246, 2003.

[195] J. C. Spall. *Introduction to Stochastic Search and Optimization*. John Wiley and Sons, Hoboken, NJ, 2003.

[196] D. Srinivasagupta, H. Schattler, and B. Joseph. Time-stamped model predictive control: an algorithm for control of processes with random delays. *Computers and Chemical Engineering*, 28(8):1337–1346, 2004.

[197] X. Su, P. Shi, L. Wu, and M. V. Basin. Reliable filtering with strict dissipativity for T-S fuzzy time-delay systems. *IEEE Transactions on Cybernetics*, 44(12):2470–2483, 2014.

[198] Y. Tang, H. Gao, W. Zhang, and J. Kurths. Leader-following consensus of a class of stochastic delayed multi-agent systems with partial mixed impulses. *Automatica*, 53:346–354, 2015.

[199] T. Tarn and Y. Rasis. Observers for nonlinear stochastic systems. *IEEE Transactions on Automatic Control*, 21(6):441–447, 1976.

[200] A. Teixeira, K. C. Sou, H. Sandberg, and K. H. Johansson. Secure control systems: a quantitative risk management approach. *IEEE Control Systems Magazine*, 35(1):24–45, 2015.

[201] R. Tempo. Robust estimation and filtering in the presence of bounded noise. *IEEE Transactions on Automatic Control*, 33(9):864–867, 1988.

[202] M. H. Terra, J. Y. Ishihara, G. Jesus, and J. P. Cerri. Robust estimation for discrete-time Markovian jump linear systems. *IEEE Transactions on Automatic Control*, 58(8):2065–2071, 2013.

[203] Y. P. Tian, S. Zhong, and Q. Cao. Structural modeling and convergence analysis of consensus-based time synchronization algorithms over networks: Non-topological conditions. *Automatica*, 65(3):64–75, 2016.

[204] Y. Toshiyuki, S. Nakatani, A. Adachi, and K. Ohkura. Adaptive role assignment for self-organized flocking of a real robotic swarm. *Artificial Life and Robotics*, 21(4):405–410, 2016.

[205] R. Tron, J. Thomas, G. Loianno, K. Daniilidis, and V. Kumar. A distributed optimization framework for localization and formation control: Applications to vision-based measurements. *IEEE Control Systems Magazine*, 36(4):22–44, 2016.

[206] V. Ugrinovskii. Distributed robust filtering with H_∞ consensus of estimates. *Automatica*, 47(1):1–13, 2011.

[207] V. Ugrinovskii. Distributed robust estimation over randomly switching networks using H_∞ consensus. *Automatica*, 49(1):160–168, 2013.

[208] V. Utkin. Variable structure control systems with sliding mode. *IEEE Transactions on Automatic Control*, 22:212–222, 1977.

[209] B. van. den. Broek, W. Wiegerinck, and B. Kappen. Graphical model inference in optimal control of stochastic multi-agent systems. *Journal of Artificial Intelligence Research*, 32:95–122, 2008.

[210] S. Varma and K. Kumar. Fault tolerant satellite attitude control using solar radiation pressure based on nonlinear adaptive sliding mode. *Acta Astronautica*, 66:486–500, 2010.

[211] A. Vempaty, O. Ozdemir, K. Agrawal, H. Chen, and P. K. Varshney. Localization in wireless sensor networks: Byzantines and mitigation techniques. *IEEE Transactions on Signal Processing*, 61(6):1495–1508, 2013.

[212] Y. Wang, W. Gao and A. Homaifa. Discrete-time variable structure control systems. *IEEE Transactions on Industrial Electronics*, 42(2):117–122, 1995.

[213] G. C. Walsh, H. Ye, and L. G. Bushnell. Stability analysis of networked control systems. *IEEE Transactions on Control Systems Technology*, 10(3):438–446, 2001.

[214] Y. Wan, G. Wen, J. Cao, and W. Yu. Distributed node-to-node consensus of multi-agent systems with stochastic sampling. *International Journal of Robust and Nonlinear Control*, 26:110–124, 2016.

[215] B. C. Wang and J. F. Zhang. Distributed output feedback control of Markov jump multi-agent systems. *Automatica*, 49(5):1397–1402, 2013.

[216] J. Wang and N. Elia. Mitigation of complex behavior over networked systems: Analysis of spatially invariant structures. *Automatica*, 49:1626–1638, 2013.

[217] L. Wang, Z. Wang, T. Huang, and G. Wei. An event-triggered approach to state estimation for a class of complex networks with mixed time delays and nonlinearities. *IEEE Transactions on Cybernetics*, 46(11):2497–2508, 2016.

[218] W. Wang and K. Chang. Variable structure-based covariance assignment for sotchastic multivariable model reference systems. *Automatica*, 36:141–146, 2000.

[219] Z. Wang, D. W. C. Ho, and X. Liu. Robust filtering under randomly varying sensor delay with variance constraints. *IEEE Transactions on Circuits and Systems II*, 51(6):320–326, 2004.

[220] Z. Wang, F. Yang, D. W. C. Ho, and X. Liu. Robust H_∞ filtering for stochastic time-delay systems with missing measurements. *IEEE Transactions on Signal Processing*, 54(7):2579–2587, 2006.

[221] G. Wei, L. Wang, and Y. Liu. H_∞ control for a class of multi-agent systems via a stochastic sampled-data method. *IET Control Theory and Applications*, 9(14):2057–2065, 2015.

[222] G. Wei, Z. Wang, and H. Shu. Robust filtering with stochastic nonlinearities and multiple missing measurements. *Automatica*, 45:836–841, 2009.

[223] N. Wiener. *The extrapolation, interpolation and smoothing of stationary time series with engineering applications*. Wiley, New York, 1949.

[224] H. S. Witsenhausen. Sets of possible states of linear systems given perturbed observations. *IEEE Transactions on Automatic Control*, 13(5):556–558, 1967.

[225] K. Won, X. Rong, and J. Wu. Multi-agent consensus control under Markovian switching topology and time-delay. *Applied Mechanics and Materials*, 427–429:750–4, 2013.

[226] J. Wu and Y. Shi. Consensus in multi-agent systems with random delays governed by a Markov chain. *Systems and Control Letters*, 60(7):863–870, 2011.

[227] X. Wu, Y. Tang, J. Cao, and W. Zhang. Distributed consensus of stochastic delayed multi-agent systems under asynchronous switching. *IEEE Transactions on Cybernetics*, 46(8):1817–1827, 2016.

[228] X. Wu, Y. Tang, and W. Zhang. Stability analysis of stochastic delayed systems with an application to multi-agent systems. *IEEE Transactions on Automatic Control*, 61(12):4143–4149, 2016.

[229] Y. Wu and L. Wang. Sampled-data consensus for multi-agent systems with quantised communication. *International Journal of Control*, 88(2):413–428, 2015.

[230] Y. Xia, G.P. Liu, P. Shi, Chen J, D. Rees, and J. Liang. Sliding mode control of uncertain linear discrete time systems with input delay. *IET Control Theory and Applications*, 1(4):1169–1175, 2007.

[231] T. Xie, X. Liao, and H. Li. Leader-following consensus in second-order multi-agent systems with input time delay: An event-triggered sampling approach. *Neurocomputing*, 177:130–135, 2016.

[232] X. He, Y. Liu, Z. Wang and D. H. Zhou. Finite-horizon quantized H_∞ filter design for a class of time-varying systems under event-triggered transmissions. *Systems and Control Letters*, 103:38–44, 2017.

[233] M. Yan and Y. Shi. Robust discrete-time sliding mode control for uncertain systems with time-varying state delay. *IET Control Theory and Applications*, 2(8):662–674, 2008.

[234] F. Yang, H. Dong, Z. Wang, W. Ren, and F. E. Alsaadi. A new approach to non-fragile state estimation for continuous neural networks with time-delays. *Neurocomputing*, 197:205–211, 2016.

[235] F. Yang and Y. Li. Set-membership filtering for systems with sensor saturation. *Automatica*, 4:1896–1902, 2009.

[236] F. Yang, Z. Wang, Y. Hung, and M. Gani. H_∞ control for networked systems with random communication delays. *IEEE Transactions on Automatic Control*, 51(3):511–518, 2006.

[237] F. Yang, Z. Wang, and Y. S. Hung. Robust Kalman filtering for discrete time-varying uncertain systems with multiplicative noises. *IEEE Transactions on Automatic Control*, 47(7):1179–1183, 2002.

[238] G. Yang and D. Ye. Adaptive reliable $\mathcal{H}_\infty$ filtering against sensor failures. *IEEE Transactions on Signal Processing*, 55(7):3161–3171, 2007.

[239] E. Yaz. A control scheme for a class of discrete nonlinear stochastic systems. *IEEE Transactions on Automatic Control*, 32(1):77–80, 1987.

[240] Y. Yaz and E. Yaz. On LMI formulations of some problems arising in nonlinear stochastic system analysis. *IEEE Transactions on Automatic Control*, 44(4):813–816, 1999.

[241] J. W. Yi, Y. W. Wang, and J. W. Xiao. Consensus in Markovian jump second-order multi-agent systems with random communication delay. *IET Control Theory and Applications*, 8(16):1666–1675, 2014.

[242] J. W. Yi, Y. W. Wang, J. W. Xiao, and Y. Chen. Consensus in second-order Markovian jump multi-agent systems via impulsive control using sampled information with heterogenous delays. *Asian Journal of Control*, 18(5):1940–1949, 2016.

[243] X. Yin and D. Yue. Event-triggered tracking control for heterogeneous multi-agent systems with Markov communication delays. *Journal of the Franklin Institute Engineering and Applied Mathematics*, 350(5):1312–1334, 2013.

[244] K. You and L. Xie. Coordination of discrete-time multi-agent systems via relative output feedback. *International Journal of Robust and Nonlinear Control*, 21(13):1587–1605, 2011.

[245] Y. Yu, H. Dong, Z. Wang, W. Ren, and F. E. Alsaadi. Design of non-fragile state estimators for discrete time-delayed neural networks with parameter uncertainties. *Neurocomputing*, 182:18–24, 2016.

[246] F. Yuan, Y. Yang, and Y. Zhang. Sampling-based event-triggered consensus for multi-agent systems. *Neurocomputing*, 191:141–147, 2016.

[247] B. Zhang and W. He. Sampled-data consensus of nonlinear multi-agent systems with stochastic disturbances. In *Proceedings of the 41st Annual Conference of the IEEE Industrial Electronics Society, Yokohama, Japan*, pages 2207–2212, 2015.

[248] D. Zhang, L. Yu, and W. A. Zhang. Energy efficient distributed filtering for a class of nonlinear systems in sensor networks. *IEEE Sensors Journal*, 15(5):3026–3036, 2015.

[249] J. Zhang, L. Ma, and Y. Liu. Passivity analysis for discrete-time neural networks with mixed time-delays and randomly occurring quantization effects. *Neurocomputing*, 216:657–665, 2016.

[250] L. Zhang, Y. Shi, T. Chen, and B. Huang. A new method for stabilization of networked control systems with random delays. *IEEE Transactions on Automatic Control*, 50(8):1177–1181, 2005.

[251] Q. Zhang and J. F. Zhang. Distributed consensus of continuous-time multi-agent systems with Markovian switching topologies and stochastic communication noises. *Journal of Systems Science and Mathematical Sciences*, 31(9):1097–1110, 2011.

[252] Q. Zhang and J. F. Zhang. Adaptive tracking games for coupled stochastic linear multi-agent systems: Stability, optimality and robustness. *IEEE Transactions on Automatic Control*, 58(11):2862–2877, 2013.

[253] S. Zhang, Z. Wang, D. Ding, F. E. Alsaadi H. Dong, and T. Hayat. Nonfragile H_∞ fuzzy filtering with randomly occurring gain variations and channel fadings. *IEEE Transactions on Fuzzy Systems*, 24(3):505–518, 2016.

[254] S. Zhang, Z. Wang, D. Ding, and H. Shu. Fuzzy filtering with randomly occurring parameter uncertainties, interval delays, and channel fadings. *IEEE Transactions on Cybernetics*, 44(3):406–417, 2014.

[255] W. Zhang, M. Branicky, and S. M. Phillips. Stability of networked control systems. *IEEE Control Systems Magazine*, 21(1):84–99, 2001.

[256] W. Zhang, Z. Wang, Y. Liu, D. Ding, and F. E. Alsaadi. Event-based state estimation for a class of complex networks with time-varying delays: A comparison principle approach. *Physics Letters A*, 381(1):10–18, 2017.

[257] X. Zhang, L. Lu, and G. Feng. Leader-follower consensus of time-varying nonlinear multi-agent systems. *Automatica*, 52:8–14, 2015.

[258] X. Zhang and J. Zhang. Sampled-data consensus of general linear multi-agent systems under switching topologies: Averaging method. *International Journal of Control*, 90(2):291–304, 2017.

[259] Y. Zhang, Z. Wang, and L. Ma. Variance-constrained state estimation for networked multi-rate systems with measurement quantization and probabilistic sensor failures. *International Journal of Robust and Nonlinear Control*, 26(16):3439–3670, 2016.

[260] Y. Zhang and W. Zhang. Guaranteed cost consensus protocol design for linear multi-agent systems with sampled-data information: An input delay approach. *ISA Transactions*, 67(0):87–97, 2017.

[261] B. Zhao, B. Xian, Y. Zhang, and X. Zhang. Nonlinear Robust Adaptive Tracking Control of a Quadrotor UAV Via Immersion and Invariance Methodology. *IEEE Transactions on Industrial Electronics*, 62(5):2891–2902, 2015.

[262] H. Zhao, S. Xu, and D. Yuan. Consensus of data-sampled multi-agent systems with Markovian switching topologies. *Asian Journal of Control*, 14(5):1366–1373, 2012.

[263] H. Zhao, S. Xu, D. Yuan, J. Lu, and Y. Zou. Minimum communication cost consensus in multi-agent systems with Markov chain patterns. *IET Control Theory and Applications*, 5(1):63–68, 2011.

[264] L. Zhao and Y. Jia. Finite-time consensus for second-order stochastic multi-agent systems with nonlinear dynamics. *Applied Mathematics and Computation*, 270:278–290, 2015.

[265] Y. Zheng, W. Chen, and L. Wang. Finite-time consensus for stochastic multi-agent systems. *International Journal of Control*, 84(10):1644–1652, 2011.

[266] Y. Zheng, G.M. Dimirovski, Y. Jing, and M. Yang. Discrete-time sliding mode control of nonlinear systems. In *Proceedings of the2007 American Control Conference, New York, USA*, pages 3825–3830, 2007.

[267] Q. Zhu and T. Başar. Game-theoretic methods for robustness, security, and resilience of cyberphysical control systems: games-in-games principle for optimal cross-layer resilient control systems. *IEEE Control Systems Magazine*, 35(1):46–65, 2015.

[268] S. Y. Zhu, C. L. Chen, and X. P. Guan. Consensus protocol for heterogeneous multi-agent systems: A Markov chain approach. *Chinese Physics B*, 22(1):Article ID 018901, 2013.

[269] X. Zong, T. Li, and J. F. Zhang. Stochastic consensus of continuous time multi-agent systems with additive measurement noises. In *Proceedings of the 54th IEEE Conference on Decision and Control, Osaka, Japan*, pages 543–548, 2015.

[270] X. Zong, T. Li, and J. F. Zhang. Stochastic consensus of linear multi-agent systems with multiplicative measurement noises. In *Proceedings of the 12th IEEE International Conference on Control and Automation, Kathmandu, Nepal*, pages 7–12, 2016.

[271] L. Zou, Z. Wang, and H. Gao. Set-membership filtering for time-varying systems with mixed time-delays under Round-Robin and Weighted Try-Once-Discard protocols. *Automatica*, 74:341–348, 2016.

[272] L. Zou, Z. Wang, H. Gao, and X. Liu. Event-triggered state estimation for complex networks with mixed time delays via sampled data information: The continuous-time case. *IEEE Transactions on Cybernetics*, 45(12):2804–2815, 2015.

[273] L. Zou, Z. Wang, H. Gao, and X. Liu. State estimation for discrete-time dynamical networks with time-varying delays and stochastic disturbances under the Round-Robin protocol. *IEEE Transactions on Neural Networks and Learning Systems*, 28(5):1139–1151, 2017.

Index

Milton Keynes UK
Ingram Content Group UK Ltd.
UKHW040104071024
449327UK00019B/788

9 780367 656867